# Introduction to Statistics for Biology

## THIRD EDITION

# Introduction to Statistics for Biology

## THIRD EDITION

Robin H. McCleery

Trudy A. Watt

Tom Hart

Chapman & Hall/CRC
Taylor & Francis Group
Boca Raton   London   New York

Chapman & Hall/CRC is an imprint of the
Taylor & Francis Group, an informa business

Chapman & Hall/CRC
Taylor & Francis Group
6000 Broken Sound Parkway NW, Suite 300
Boca Raton, FL 33487-2742

© 2007 by Taylor & Francis Group, LLC
Chapman & Hall/CRC is an imprint of Taylor & Francis Group, an Informa business

No claim to original U.S. Government works
Printed in the United States of America on acid-free paper
10 9 8 7 6 5 4 3 2 1

International Standard Book Number-10: 1-58488-652-8 (Softcover)
International Standard Book Number-13: 978-1-58488-652-5 (Softcover)

**Visit the Taylor & Francis Web site at**
**http://www.taylorandfrancis.com**

**and the CRC Press Web site at**
**http://www.crcpress.com**

*To Colyear Dawkins*

# Contributors

**Dr. Robin McCleery,** Edward Grey Institute, Department of Zoology, University of Oxford, South Parks Road, Oxford, OX1 3PS, Email: robin.mccleery@zoo.ox.ac.uk, Tel 01865 271161, Fax 01865 271168.

**Dr. Trudy A. Watt,** Trinity College, Broad Street, Oxford, OX1 3BH, Email: trudy.watt@trinity.ox.ac.uk, Tel 01865 279881, Fax 01865 279911.

**Mr. Tom Hart,** 1. Imperial College at Silwood Park, Ascot, Berks, SL5 7PY, Email: tom.hart@imperial.ac.uk, Tel. 0207 594 2447, (no fax); 2. British Antarctic Survey, High Cross, Madingley Road, Cambridge, CB3 0ET, Tel: 0207 594 2307.

# Contents

# *Preface*

Students of biological subjects in further and higher education often lack confidence in their numerical abilities. Those coming straight from school are likely to have some science subjects in their school leaving qualifications but may have little mathematics. Mature students often describe themselves as "rusty," and some biologists choose biological sciences specifically to escape from numbers. Unfortunately for them, an understanding of experimental design and statistics are as central to modern biology as an understanding of evolution. As well as demonstrating the importance of statistics, this text aims to calm fears and to provide a relatively painless way into the subject.

We stress a reliance on computers throughout. Although we demonstrate procedures and concepts in all tests, no one in "the real world" would conceivably do anything but the simplest statistics by hand, or rely solely on a calculator. Modern statistical programs are very good at analysing large data sets and are becoming increasingly good at reporting errors to show when a test might not be appropriate.

The title of the third edition has been shortened from *Introductory Statistics for Biology Students* to the more inclusive *Introductory Statistics for Biology*, which reflects the fact that many researchers refer back to introductory statistics texts to refresh their memories.

The second edition was published in 1997, and by 2007, a new edition was overdue. MINITAB has reached its 15th edition with the incorporation of numerous improvements, and the third edition of this book is set to coincide with its release. These changes are reflected throughout the book both in the range of tests that can be performed and increased clarity to the beginner. Robin McCleery and Tom Hart have joined Trudy Watt as coauthors.

Trudy Watt wrote the first and second editions while a senior lecturer in Statistics and Ecology at Wye College, University of London. Robin McCleery's experience of teaching first-year biology students at Oxford University with the 2nd edition as a textbook has suggested some areas for expansion, some for deletion, and a change in the balance of topics. Tom Hart was a tutor and demonstrator on this course in the Zoology Department, and both have tried to address issues that current students struggle with.

Although the ethos remains the same, there are some significant changes:

- Removal of many of the exercises, which have been replaced by worked examples.
- A new general template for carrying out statistical tests from hypothesis to interpretation. We repeat this throughout to show the generality to different types of tests.

- An emphasis on experimental design, and simulating data prior to carrying out an experiment.
- MINITAB analyses and graphics have been updated to Releases 14 and 15.

In particular, the changes are:

Chapters 1 to 3 have been greatly expanded to provide a more thorough grounding in the basic ideas behind statistical thinking. We explain probability in some detail to clarify the rationale behind hypothesis testing before moving on to simple tests. We have emphasised the common thread running through much of statistics by formulating a general approach to carrying out a statistical test. Some more detailed explanation in the early chapters ensures that we can follow the same template throughout the book.

Chapter 4 represents the chapters on sampling and experimental design from the 2nd edition, which we have combined for simplicity and because they are so intrinsically linked.

Chapters 5 to 7 on Analysis of Variance (ANOVA) are very similar in scope to previous editions but have been rewritten with more emphasis on factorial designs and interactions. We have also removed many references to post-hoc testing to discourage people from excessive use of these tests.

Chapter 8 (Correlation and Regression) has been rewritten to bring it more into line with the approach taken in Chapters 5 to 7, where we initially introduce the method with a very simple numerical example. We have also made the similarities between ANOVA and regression more apparent.

The chapter concerning data from an observational study has been deleted to allow for a more thorough discussion of categorical data and nonparametric statistics in Chapter 9 and Chapter 10. What was previously discussed in the chapter on observational studies has now been partially covered in Chapters 9 to 11, and partly covered by references.

Chapter 11 is now a general project template with advice on how to carry out and write up an undergraduate project. We also include a short sample report to illustrate many of the points we make.

We have included a trial copy of MINITAB version 15 for you to try. Just insert the disk and follow the instructions to get a 30 day free trial. Details of purchase can be found at http://www.minitab.com.

MINITAB has a large number of data sets available for practice use. Descriptions are found by consulting the index, selecting data sets, and clicking on the name of one. To upload a data set into the worksheet, click on "file, open worksheet" and select the desired filename. The reader is encouraged to use these data sets, which replace the end-of-chapter examples in the 2nd edition. For example:

POTATO.MTW. In this experiment a rot-causing bacterium was injected into potatoes in low, medium, or high amounts (C1). The potatoes were left for 5 days at 10 or 16 degrees C (C2) and with 2, 6, or 10% oxygen (C3). The diameter of the rotted area on each potato was measured as the response variable (C4).

YIELDSTDV.MTW. In this study, there are eight blocks (C4) and three factors: reaction time (C5), reaction temperature (C6), and catalyst (C7). The yield of the chemical reaction was recorded (C8).

Our inspirations and interest in statistics are varied, but this book remains dedicated to the late Colyear Dawkins whose enthusiasm for communicating the principles of experimental design and analysis was infectious. We would also like to thank a number of people for comments while writing this new edition, in particular, Marian Dawkins, Marta Szulkin, Matt Towers and Liz Masden for comments on the text.

---

## Note to Students

Statistics is an essential tool for all life scientists. One of the most important parts of a college or university course in biology and related subjects is learning to understand and critically evaluate research evidence. Open any scientific journal in the life sciences and you will find the pages littered with probability statements, test statistics, and other jargon that must be understood if you are to make sense of the claims being made by the researchers. Also, as an apprentice scientist yourself, you will soon start to undertake your own investigations and need the right tools for the correct interpretation of the results.

Unfortunately, a first-year statistics course is often seen as just an inconvenient hurdle to be jumped, especially by students who may have little mathematical background. This tends to make people focus on solving problems, which they often do by rote and without perceiving the underlying rationale. In our view this approach actually makes it harder to understand the subject, and we feel that you really need to develop some curiosity about the why's and wherefore's as well as the "how to." One result of this approach is that you do need to read the book from start to finish, rather than dipping in for the bit you want. The argument builds, and we have not always cross referred back to the basic material at every stage.

Many introductory texts contain a disclaimer about the mathematics they will require you to tackle but promptly renege on their promise to shield you from the horrors of equations within a page or two. The fact is that, without some understanding of equations, statistics is going to be hard to explain, but it is also true that "serious mathematics," by which we mean an understanding of techniques such as calculus, the proof of theorems, and so on, is not really necessary. Most of the ideas involved can be given an intuitive or visual representation, and that is the approach we use in this book.

Another feature of the book is a reliance on the use of computerised methods. There is a tradition of sadism amongst teachers, matched by masochism in some students, who share the idea that unless you have experienced the agony of manual calculation you do not really understand the subject. We dissent from this view for a number of reasons. For some people the effort involved in getting the calculations right gets in the way of understanding. Many manual methods employ algebraic shortcuts to reduce the amount of key pressing on the calculator, but without some facility at algebra, these completely obscure the underlying rationale. It is, of course, true that computers can make things a bit too easy. The numbers go in one end and a printout comes out of the other, but the user may not really have much of a clue what it all means. What we aim to do here is to explain the ideas behind the methods and let the calculating machinery do the drudgery, but you do need to be curious about all the numbers that appear in the output. Most of them are there for a reason, and you should remain uncomfortable until you know what each of them represents.

Although generic computer programs such as spreadsheets contain statistical functions, they often have shortcomings in reliability and in the ease with which statistical information is presented. We have chosen to emphasise the use of specialised statistical packages, and specifically one called MINITAB. Almost everything in this book can be achieved using the student version of this program (Student 14) which is modestly priced.[1] The full release of MINITAB is currently Release 15, and it has a number of facilities not found in the Student release. We have included a time-limited trial copy on CD for you to try. MINITAB has much to recommend it for this purpose, as it was originally written primarily as a teaching tool. It incorporates many useful functions for exploring the ideas behind conventional tests, and contains built-in tutorial, reference, and help materials. Mainstream packages (the full version of MINITAB, SPSS, SAS, Genstat) are expensive for individuals to purchase, though many institutions have site licences that allow registered students to use them at a much lower cost. In practice, there is a high degree of convergence in the way the user interfaces operate, so the skills learned using MINITAB are readily transferred.

Recently, there has been another innovation, the existence of a free statistical package called R (see http://cran.r-project.org). The main problem with this package, in our view, is the user interface, which relies almost entirely on commands typed in from the keyboard. Our experiences teaching with early releases of MINITAB, which worked in a similar way, convince us that most people find the "point and click" interface much easier. However,

---

[1] MINITAB is a registered trademark of Minitab Inc., whose cooperation we gratefully acknowledge http://www.minitab.com/. Student 14 release details: http://www.minitab.com/products/minitab/student/default.aspx.

MINITAB can help prepare you for something like R because the MINITAB command line lives on behind the scenes (see Appendix A, Section A.10).

Most beginners in statistics find that "choosing the right test" is the hardest bit of the subject and are looking for a "cookbook" or perhaps a taxonomic key to tell them what procedure to follow in a given case. In fact, even experienced statisticians will admit that they sometimes see how to solve a problem by spotting an analogy with an analysis they have encountered elsewhere in a book or journal. We could have written the book in the form of a series of recipes (and we present a taxonomic key in Appendix E) but in our view this would not achieve our objective, which is to encourage you to think statistically. Experienced cooks rarely use a recipe book, except for inspiration, because they have realised that there is only a small number of basic techniques to know. Gravy, béchamel and cheese sauce are variants of a single method; what you need to understand is how to combine a fat, a thickening agent and a liquid without making it lumpy. Similarly, once you start to get the hang of statistical thinking, you realise that there is a single thread running through the whole subject. What at first sight seems to be a book full of disparate tests (many with impressive-sounding names) is a consistent series of methods all based on a common underlying structure of reasoning. Our hope is that, by the time you have finished using this book, you will begin to appreciate this unity in statistical thought and, maybe, just a little, start to share our fascination with it.

### Conventions in Our Presentation

Generally, concepts appearing for the first time appear in **bold** and emphasis is made using *italics*. We have also used bold and normal fonts to represent input to MINITAB, as explained in Appendix A.

We have not thought it necessary to number equations, with a few obvious exceptions.

The results of manual calculations are generally given to 2 places of decimals (2 d.p.) rounded off in the usual way. Thus, 2.561 becomes 2.56 (to 2 d.p.), and 2.565 becomes 2.57.

Numbers less than 0.01 are usually given to 3 significant figures. If you see $p = 0.000$ in a computer output, it is unlikely that the probability is really 0! It means $p < 0.0005$.

Some care must be taken when verifying calculations manually using intermediate numbers printed by the computer. The computer often rounds off its output but will do any further calculations using the full numerical values and rounding off the result. This can give rise to apparent rounding errors in some circumstances, but we think we have found all the cases where this might cause confusion.

# 1

## How Long Is a Worm?

> I would not enter on my list of friends ... the man who needlessly sets foot upon a worm.
>
> **—William Cowper**

## 1.1 Introduction

School experiments in physics and chemistry often have known answers. If you do not record a value of 9.8 m s$^{-1}$ s$^{-1}$ for "the acceleration with which an object falls to the earth," then you know it must be because there was something wrong with your equipment or with how you used it. Similarly, the molar mass of calcium carbonate is 100.09, so any other value would be wrong. The idea that there is a single clear-cut answer to a question is frequently not relevant in biology. Questions such as "How heavy are hedgehogs?" or "What is the length of an earthworm?" do not have a single correct answer. In fact, the variation may be of biological interest in itself. For example, the weights of individual hedgehogs in autumn largely determine whether they will survive their over-winter hibernation. It is thus important to be able to give a precise answer to such questions, and the aim of this chapter is to show how to do it correctly. We will simplify life by concentrating on just those earthworms of one species living in one particular field. Because earthworms are both male and female at the same time, we do not need to specify which sex we wish to measure. Even so, individuals can have one of a wide range of lengths. Why is this? Earthworms can live for a long time, and young earthworms are likely to be shorter than old ones. Like all animals, earthworms have genetic variability — some have a tendency to be short and fat and others to be long and thin. Those that live in the moister part of the field at the bottom of the slope might be more active and have a better food supply, so they will grow more quickly and may tend to be longer than those in a less favourable part of the field. Meanwhile, those living near the footpath along one side of the field tend to be shorter because they are infested with a parasite or because they have recently escaped from a tussle with a bird. How then should we measure and describe the length of worms in this field?

---

## 1.2   Sampling a Population

We would like to obtain information about *all* the worms in the field because they form our **population** of interest. It is difficult or impossible to measure them all, so we will measure only a few — our **sample** — and generalise our results to the whole **population**. Because we want to draw inferences about the whole population on the basis of our sample, it is important that the worms in the sample are *representative* of it. For example, if we wanted to make inferences about earthworms in the whole of Oxfordshire, we would sample from more than just one field because we could not expect the conditions in a single field (and hence the properties of the worms in it) to be typical of the whole county.

### 1.2.1   Measuring Worms

Imagine that we have collected ten worms and propose to measure them using a ruler. There are practical limitations to the accuracy or our measurement. Worms wriggle and stretch and contract. We will need to make them keep straight and still, and to coax them into a straight line with one end at 0 mm while we read off the measurement. We must avoid stretching the worm or allowing it to contract too much. Doing this consistently is difficult and will need practice. Although we would like to say that "This particular worm is 83 mm long" and mean it, we will realise that this is unlikely to be a very accurate measurement. We can only do our best to standardise the process, and it may be that we decide to measure earthworms only to the nearest 5 mm, i.e., 0.5 cm. (see Box 1.1).

---

### Box 1.1 Measurement Errors

Of course, we have been a bit optimistic about the business of making scientific measurements. If a class of students tried to measure the gravitation constant (page 1), it is unlikely they would all get the same answer of $9.8 \text{ m s}^{-1} \text{ s}^{-1}$ even if they followed the correct procedures. The reason for this is *measurement error*. When you read a ruler or start and stop a clock, there is an unavoidable error due to your eyesight, your reflexes, and so on. However, it is true that if we took the average of all the class measurements, it would almost always be very close to $9.8 \text{ m s}^{-1} \text{ s}^{-1}$ because, on average, all the measurement errors would cancel out. If we all try to measure the lengths of a set of earthworm specimens, there are actually two sources of variation — the measurement errors due to our varying competences with a ruler, and the natural variability amongst earthworms.

Here are our results in centimetres:

11.5, 10.0, 9.5, 8.0, 12.5, 13.5, 9.5, 10.5, 9.0, 6.0

## 1.2.2 Summary Measures: "Centre"

How can we summarise the information represented by these numbers? We could just draw a picture of it (Figure 1.1a). This graph is called a **dot plot**; you draw a horizontal line and mark out the possible lengths, and then place a dot for each worm at the point on the line representing that particular length. Immediately, we see quite a spread of lengths, and that there seem to be more in the middle than at the sides. A "typical" worm from this sample then would be one of the **average** or **mean** length. If we had to guess how long one worm in the sample was, we would minimise the difference between our guess and the actual worm's length by guessing that it was the mean length. If we add up the ten lengths and divide the sum by ten, we get 10 cm, which is the average or mean length of worms in the sample (the **sample mean**).

This is not the only possibility as a measure of the typical worm; we could instead take the commonest value (the **mode**). In this case, there are two worms of 9.5 cm and only one of all the others, so 9.5 cm is the mode — not terribly useful here, so we will not pursue this one any further.

Another possibility would be to use the value such that half the worms lie above it and half below. If we arrange our measurements in order, we can see that the midpoint is halfway between the 5th and 6th number from the left:

6.0, 8.0, 9.0, 9.5, 9.5, 10.0, 10.5, 11.5, 12.5, 13.5

Conventionally, we split the difference between 9.5 and 10 to get 9.75, which is the length of the "middle-ranking" worm, a reasonable thing to use as the central value. This measure is called the **median** length and has advantages over the mean in some circumstances (though here it is almost the same as the mean, so it makes little difference which we regard as representing the typical worm). We will defer thinking about this until Chapter 10.

## 1.2.3 Summary Measures: "Spread"

The other thing that we see more readily when we sort the data is the enormous variation in length. Some of the worms are twice as long as others.

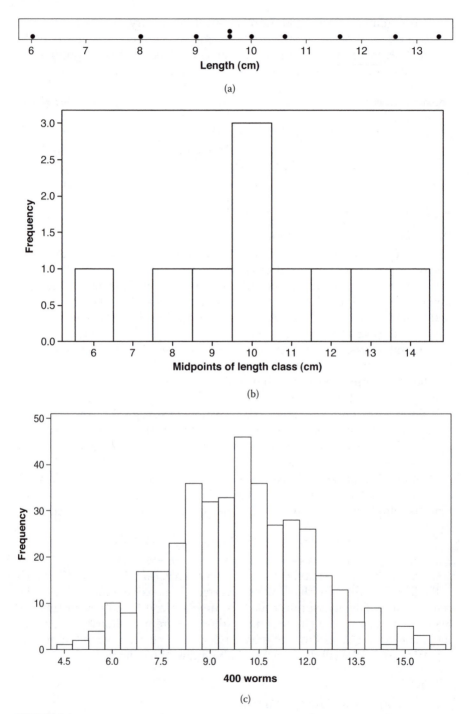

**FIGURE 1.1**
(a) Dot plot of lengths from a sample of ten worms. (b) The number of worms in each size class from a sample of ten worms. (c) The number of worms in each size class from a sample of 400 worms.

It is sensible to check our set of observations at this stage. For example, if one of them was 70 cm, do we remember measuring this giant (you do find these, but only in Australia), or is it more likely that we have misread 10 cm as 70 cm? Once we have sorted this problem out, we still have to deal with the fact that not all worms are the same length. How can we express the amount of variation in our sample?

An obvious possibility here would be to consider the **range** of values in the sample — in this case, the measurements go from 6.0 to 13.5 cm, spanning a range of 7.5 cm. A problem with the range is that it gives the same importance to the ends of the distribution as it does to the middle, failing to capture the fact that more of the measurements are in the middle. A better bet might be the interval within which half the measurements lie. This is called the **interquartile range** and is the distance from the length below which a quarter of the measurements lie (called the **lower quartile** for obvious reasons) and the length above which a quarter of the measurements lie (**upper quartile**). In this case, the lower quartile is 8.75 and the upper quartile is 11.75, so the interquartile range is 3 cm (the precise calculations for this are given in Section 10.2).

However, the measure of spread that we favour for this type of continuous measurement data is something called the **standard deviation** (s.d.).

Calculating the sample standard deviation is a bit more complicated than the interquartile range. We give an outline of the method here but we will justify it in detail, in Section 1.6.4. The procedure is:

1. Find the mean.
2. Subtract the mean from each data point to get its deviation from the mean.
3. Square the deviations from the mean and add them up.
4. Divide by one fewer than the number of data points: this gives the **sample variance**.
5. Take the square root to get the **sample standard deviation**.

The answer, for our sample of worms, is 2.173 (see Box 1.2).

---

## Box 1.2 Coming to Terms with Those Equations

Unlike many introductory texts, we did not promise to shield you from equations, but we did say we would try to explain how they work. Here we build up one of the most important equations you have to deal with — that for the calculation of the sample standard deviation. We go through the rationale for this later in Section 1.6.4.

The numbers refer to those in the procedure first mentioned in this section.

1. Find the mean. We have $n$ observations. We call each one $x_i$ where the subscript $i$ means we are dealing with the $i$-th observation. To get their sum we will write:

$$sum(x) = \sum_{i=1}^{n} x_i,$$

   meaning "set $i$ to each possible value and add the $i$-th $x$-value to the total"

Thus, the mathematical formula for the mean is

$$\bar{x} = \frac{\sum_{i=1}^{n} x_i}{n}.$$

We write $\bar{x}$ for the mean, pronouncing it "x-bar."

2. Subtract the mean from each data point. So, for each value of $i$ from 1 to $n$, we calculate

$$Deviation\ from\ the\ mean = (x_i - \bar{x})$$

3. Square the deviations from the mean and add them all up. Using the same summation idea as in step 1, we calculate

$$Sum\ of\ squared\ deviations\ about\ the\ mean = \sum_{i=1}^{n} (x_i - \bar{x})^2$$

4. Divide by one fewer than the number of data points to get the variance

$$Sample\ variance = \frac{\sum_{i=1}^{n} (x_i - \bar{x})^2}{(n-1)}$$

5. Take the square root to get the standard deviation

$$Sample\ standard\ deviation = \sqrt{\frac{\sum_{i=1}^{n} (x_i - \bar{x})^2}{(n-1)}}$$

In practice, it is common to see this formula without the $i$ subscripts:

$$sd = \sqrt{\frac{\sum (x - \bar{x})^2}{(n-1)}}$$

where a bare $x$ term means "one of the $n$ measurements."

This is the definitional formula: you would not normally use this method to calculate it! Use a calculator or a computer. Do not forget that on a calculator you must use the button for the sample standard deviation to get this result. It normally has the symbol "$\sigma_{n-1}$" or sd$(n-1)$ on it.

Although it seems a long-winded way of getting a measure of spread, the standard deviation has certain very useful properties. The greater the standard deviation, the bigger the spread, but it is more sensitive to extreme values than the interquartile range, which just throws away information about values beyond the upper and lower quartiles. Another useful property is the fact that at least 75% of the measurements in any kind of sample are likely to lie within 2 standard deviations of the mean. This remarkable claim is called Chebyshev's rule, though it is not quite as useful as it looks. For example, "at least" is not very precise; under many circumstances the value would be nearer 96%, and we need more information about the data to tell us which figure is the right one to use. We return to this in Section 1.3.

### 1.2.4 Generalising from the Sample to the Population

So, how long are the worms in the field? Does the length of a typical worm in our sample give us any idea what the length of a typical worm in the field might be? Imagine that every worm in our sample of ten measured 10 cm. This is very consistent (not to mention very suspicious — we should check someone is not making up the results). The mean length would be 10 cm as before, but since all the worms we found were 10 cm long we would probably be very happy to conclude that the mean length of all the worms in the field (the population) is very close to this value. If it were not, it is extremely unlikely that we would pick out 10 individuals of that length. If we had sampled 20 worms and they all measured 10 cm, this would be amazingly strong evidence that the **population mean** is very close to 10 cm. The more worms we measure (increasing **replication**), the more information we have, and so the greater our confidence that the **sample mean** is close to the population mean. We often speak of the sample mean being an **estimate** of the population mean in these circumstances.

However, the ten worms in our sample were *not* all the same length. Here they are again (in order):

6.0, 8.0, 9.0, 9.5, 9.5, 10.0, 10.5, 11.5, 12.5, 13.5

The mean length is:

$$\sum x \Big/ n = (6.0 + 8.0 + ... + 13.5) \Big/ 10 = 100 \Big/ 10 = 10 \, cm$$

With this amount of variation in the ten values, how confident can we be that the mean length of *all the worms in the field* is 10 cm? And how variable are the lengths of the worms in the field likely to be, as measured by their standard deviation? To answer this question we need a way of relating the *variability* of the sample to the variability of the population.

### 1.2.5   How Reliable Is Our Sample Estimate?

#### 1.2.5.1   *Naming of Parts*

Statisticians have a special vocabulary to talk about the business of estimating the properties of a population on the basis of a sample. The **population**'s characteristic is called a **parameter**, whereas the estimate of it, obtained from the **sample**, is called a **statistic**. So, if the mean length of all the worms in the field is actually 9.8 cm, we may obtain an estimate of this *parameter* by taking a *sample* of 10 worms and calculating a *statistic*, namely the sample mean. In our example data, this sample statistic turns out to have a value of 10 cm. If we took another sample of 10 worms, we would obtain a different estimate, perhaps 9.7 cm, and so on. Because we usually only take one sample, and so have only one statistic, it is important to be able to express the reliability of our estimate of the parameter. The following sections outline how this is done.

## 1.3   The Normal Distribution

In Figure 1.1a we can see that there is a slight tendency for there to be more middle-sized than extreme worms; this becomes more noticeable as the number of worms in the sample increases (Figure 1.1c). For the whole population of worms in the field (assuming that the field does not contain any localized abnormality), the size distribution would almost always follow a smooth bell-shaped curve (Figure 1.2), called **Normal distribution**. (*Normal* is a technical word here; it does not just mean "ordinary"). The spread of the curve depends on the natural variability in the population, being greater if the variability is large. Populations of continuous measurements such as length and weight are often distributed in this way. The shape of the curve is completely defined by its mean and standard deviation.

However, not all populations will have a Normal distribution. For example, counts of the number of butterflies per thistle might produce mostly

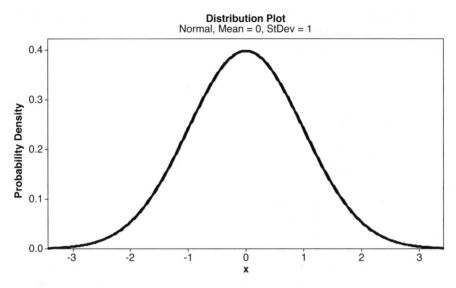

**FIGURE 1.2**
The Normal distribution.

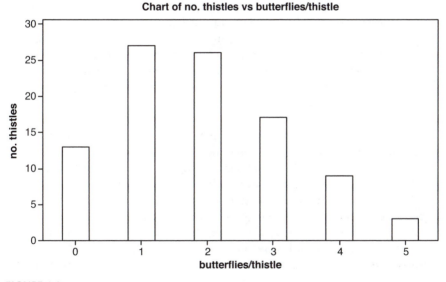

**FIGURE 1.3**
A positively skewed distribution.

zeros with a few ones and twos and the occasional larger number where the thistle happens to be in a favourable spot for butterflies. Such a distribution would be **skewed** to the right (Figure 1.3).

Although other distributions turn up in statistics (Box 1.3), most of what we have to deal with in this book depends on the properties of the Normal distribution.

## Box 1.3 Discontinuous or Discrete Distributions

The Normal distribution is **continuous** in that it is used for data
that are measurements that can take any value on a continuous scale
(e.g., mm, kg). However, data may also be counts of individual
objects or events, and so can only be whole numbers — i.e., they are
**discontinuous**. The distributions followed by such measurements
are often called **discrete distributions**. Two main types of discrete
distribution are relevant to counts and are often met with in biological data (for more detail see Rees, 4th edition).

- **Binomial distribution.** This is relevant when we have only two
  mutually exclusive categories such as "alive or dead" or "male
  or female." These are often expressed as proportions or percentages, e.g., percentage mortality. For example, the probability of getting just one boy in a family of 12 children depends
  only on the number of children, and the probability of any child
  being a boy. We examined this further in Section 1.4.4.

- **Poisson distribution.** This is relevant where we are counting
  independent events. For example, if we have a microscope slide
  of microorganisms marked out into graticule squares, then the
  number in each of several squares will follow the Poisson distribution, provided they are not clumped together in some way.
  To find the probability of a particular number turning up in
  one square, we need know only the average number per square
  over the whole slide.

Both these distributions assume that the individual events or
things we are counting are behaving "according to chance." Sometimes we may suspect that there is a biological reason for individuals
(such as the microorganisms on a slide) to be arranged in a pattern.
Methods of discovering whether this is happening or not are discussed in Chapter 9. However, if data are counts but the numbers
involved are fairly large, it is often reasonable and convenient to
treat them as if they followed an approximately Normal distribution.

## 1.4  Probability

We have already met the idea that at least 75% of measurements from any
type of distribution lie within 2 standard deviations of the mean. If we know
that our measurements follow a Normal distribution, we can be more precise

about this and say that 96% of the measurements lie within 2 standard deviations of the mean. We can also say that 68% lie within 1 standard deviation. If we draw the Normal curve by putting the mean and standard deviation measured from our sample into the equation, we are representing what proportion of values we *expect* to fall at various distances from the mean. We usually express this idea by saying (for example) that if the population from which the measurements come is Normally distributed, then the **probability** that a randomly chosen representative is within one standard deviation of the mean is 0.68. To make sense of this idea we first need to understand something about the concept of probability.

### 1.4.1 What Do We Mean by a Probability?

In ordinary language we often talk about the probability that our next child will be a boy, or the probability that I will pass my exams. If you are expecting a baby then (assuming you have not cheated by looking at the scan results), there is an approximately equal chance it will be a boy or a girl. What we actually mean here is that in a very large population of live births, about half would be boys and half girls. We express this by saying that the **probability** of a child being a boy is 1/2 or 0.5. A **probability** is a number between 0 and 1, where 1 represents certainty that an outcome will occur, and 0 represents certainty that it will not occur. The number 0.5 is exactly half way between and represents the case where the outcome is equally likely either to occur or not occur.

### 1.4.2 Writing Down Probabilities

The probability that an event A occurs is written p(A). For example, let X be the sex of a child. p(X = boy) is the probability of the event "the child is a boy"; this is sometimes shortened to p(boy). What is the probability that two children will both be boys? If you consider the first child, there is a 50:50 chance it will be a boy, so p(boy) = p(girl) = 0.5. If you have two children, then there are three possible outcomes: two boys, two girls, or one of each (assuming we do not mind the sequence of girls and boys).

Just by counting up the cases, you can see that "two boys" will occur in about one quarter of families of two children, so we could write p(boy,boy) = 0.25. Counting up is rather a tedious way to find the probability of the combined event, so we make use of the insight that the probability of two independent events occurring together (or one after the other) is found by just multiplying the two probabilities together.

What is the probability of one girl and one boy, regardless of the order? There are two ways this could happen, so we have to draw a diagram of this (Table 1.1). We can see that

$$p(boy, girl) = p(boy) \times p(girl) = 0.5 \times 0.5 = 0.25.$$

**TABLE 1.1**

The Four Possible Outcomes in a Family of Two Children

| Second child | Boy | Girl |
|---|---|---|
| First child: boy | Two boys | Boy then girl |
| First child: girl | Girl then boy | Two Girls |

In exactly the same way,

$$p(girl, boy) = p(girl) \times p(boy) = 0.5 \times 0.5 = 0.25.$$

To get the probability of one child of each sex regardless of order, we have to add the probabilities of the two outcomes together to get 0.5. On average, half of all families of two children should contain one girl and one boy. Again, this makes sense — the probability that one or other of two events will occur must be more than the probability that either of them occurs on its own.

This example illustrates two general rules for working out the probabilities of compound events — that is, outcomes involving more than one simple event.

### 1.4.2.1  *Multiplication Law (Product Rule)*

For two independent events A and B,

$$p(A \text{ and } B) = p(A) \times p(B)$$

Example: p(first child is a boy *and* second child is a boy) = $1/2 \times 1/2 = 1/4$

### 1.4.2.2  *Addition Law (Summation Rule)*

For two independent exclusive events A and B,

$$p(A \text{ or } B) = p(A) + p(B)$$

Example: p(child is a boy *or* child is a girl) = $1/2 + 1/2 = 1$

Note that it follows from the addition law that if the probability that an event will occur is *p(event)*, then the probability that it will *not* occur is given by

$$p(not event) = 1 - p(event) \quad \text{So, } p(girl) = 1 - p(boy).$$

We often use this result in statistical analysis.

### 1.4.3  **Important Restriction**

As given above, both rules relate to **independent** and **exclusive** events.

- Events are **independent** if the probability of one occurring is not affected in any way by whether or not the other has occurred: in theory the sex of one child has no effect on the sex of any subsequent sibling.

- Events are **exclusive** if one or other can occur but not both: if a child was a boy, then it was not a girl and *vice versa*.

If events are not independent and exclusive, we may have to deal with **conditional probabilities**. The probability of B given that A has occurred is written $p(B|A)$.

For reference, the full versions of the laws are given here:

- If A and B are not independent, then the multiplication law becomes

$$p(A \text{ then } B) = p(A) \times p(B|A)$$

Similarly, $p(B \text{ then } A) = p(B) \times p(A | B)$

- If A and B are not exclusive, then the addition law becomes

$$p(A \text{ or } B \text{ or both}) = p(A) + p(B) - p(A \text{ and } B)$$

We will ignore this complication for now.

### 1.4.4   An Example

How many brothers and sisters do you have? We have already said that, on average, the probability of a child being a girl or a boy is about equal: $p(girl) = p(boy) = 0.5$. It is also generally believed that the sex of one child does in fact have no effect on the probability that its subsequent siblings will be male or female. If both these beliefs are correct, the probability of two boys in a family of two children is

$$p(boy) \times p(boy) = 0.5^2 = 0.25$$

This makes sense because you would expect the probability of two boys in the family to be less than the probability of each child being a boy. As we have already seen in Section 1.4.2, the probability of one child of each sex is

$$p(boy) \times p(girl) = 0.5 \times 0.5 = 0.25$$

but this outcome can occur in two ways, so

$$p(boy + girl) = p(boy, girl) + p(girl, boy) = 0.5$$

Thus, we see both the product rule and the summation rule at work here. If we extend this reasoning to three children, we begin to see a pattern in the number of different ways each outcome can occur: We will write B = boy and G = girl to save space.

If we ignore the sequence, we can see that the number of ways each outcome could be obtained increases as the number of events increases (Table 1.2).

The probability that we will see two boys and a girl in a family of three children is thus the number of ways of arranging having two boys and a girl (3) multiplied by the probability of having BBG in any order:

$$p = 3 \times 0.5 \times 0.5 \times 0.5 = 0.375 \text{ or } \frac{3}{8}$$

We can see from Table 1.3 that a run of three boys in a row, or three girls in a row, is not that unlikely — about 1/8 (0.125) of families of three children should be all boys and the same proportion all girls — so a quarter of such families have children of all the same sex.

### 1.4.5   Probability Density Functions

In Figure 1.4 we plot a bar chart where the x-axis represents the number of boys in a family of three children, and the y-axis is the probability that there will be this number of boys. A graph like this is called the **probability density function (p.d.f.)** for the distribution of number of boys in families of three children. The graph gives a correct picture of these probabilities *provided that* the sexes are equally likely, and occur independently of each other within families.

**TABLE 1.2**

The Possible Outcomes in Families of 1, 2, or 3 Children

| 1 Child |     | B   |     | G   |     |
|---------|-----|-----|-----|-----|-----|
| 2 Children | BB | BG | | GG | |
|         |     | GB  |     |     |     |
| 3 Children | BBB | BBG | BGG | GGG |
|         |     | BGB | GBG |     |
|         |     | GBB | GGB |     |

**TABLE 1.3**

The Probabilities of Obtaining 0, 1, 2, or 3 Boys in Families of 1, 2, or 3 Children

| Children | P(X = 0 boys) | P(X = 1 boy) | P(X = 2 boys) | P(X = 3 boys) |
|----------|---------------|--------------|---------------|---------------|
| 1 | $1 \times 0.5 = 0.5$ | $1 \times 0.5 = 0.5$ | | |
| 2 | $1 \times 0.5 \times 0.5 = 0.25$ | $2 \times 0.5 \times 0.5 = 0.5$ | $1 \times 0.5 \times 0.5 = 0.25$ | |
| 3 | $1 \times 0.5 \times 0.5 \times 0.5 = 0.125$ | $3 \times 0.5 \times 0.5 \times 0.5 = 0.375$ | $3 \times 0.5 \times 0.5 \times 0.5 = 0.375$ | $1 \times 0.5 \times 0.5 \times 0.5 = 0.125$ |

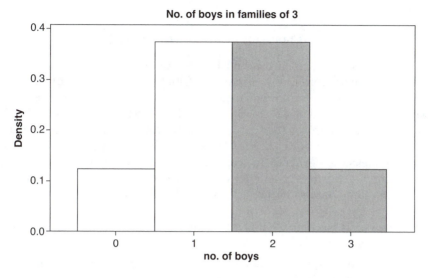

**FIGURE 1.4**
The probabilities of having a different number of boys in a family of three.

The probability of each outcome corresponds to the *area* of the relevant bar in the graph. Provided all the outcomes are included in the plot, the total area of the bars is 1.0. If we let the symbol X stand for the number of boys in a family of three children, we can write down the probability of either two or three boys as p(X = 2) + p(X = 3), which is the shaded area. The probability of fewer than two boys p(X < 2) is 1 – *shaded area*.

The unshaded area is often referred to as the **cumulative density** up to X = 1, or p(X ≤ 1). With discrete distributions (Box 1.3) like this, it just saves adding up, but the adding up idea is very important with probability distributions where the x-axis is continuous.

### 1.4.6    What Have We Drawn, Exactly?

It is very important to realise that, unlike Figure 1.1, Figure 1.4 is *not* a graph of any real data. Any particular family must have a particular number of boys and girls. What the graph shows is what proportion of a large number of families containing 3 children *would have* a specified number of boys *if we were to go out and collect the data*. It is thus a theoretical distribution, called the **Binomial distribution**, for the sample statistic "number of boys in families of 3." The Binomial distribution is completely defined by the number of independent events (3 births in this case) and the probability that each will be a boy (0.5 in this case), so we speak of "the Binomial distribution for n = 3 and p = 0.5." The graph underlines the fact that, theoretically, half the children would be boys. For families of 3, the *expected* number is a fraction — an average of n × p = 3 × 0.5 = 1.5 boys — even though real families only have whole numbers of children in them.

## 1.5   Continuous Measurements — Worms Again

On the whole, children are indivisible, so things like counts of discrete individuals tend to follow a discrete distribution such as the Binomial or Poisson. What about continuous measurements? One of the 10 worms in our sample was recorded as 6.0 cm. So, does this mean that the probability of a worm having length = 6.0 cm is $1/10 = 0.1$? Not necessarily. Do we really mean 6.001 or 5.999, for example? In fact, we have already agreed that we can only measure worms to an accuracy of 0.5 cm, so what 6.0 really represents here is a worm whose length lies between 5.5 and 6.5 cm. Figure 1.5 shows the same data as Figure 1.1b but with two subtle alterations:

- The y-axis has been changed to density (strictly, **relative frequency density**)
- The x-axis has been changed to show the **cut-points** rather than the **midpoints** of the histogram bars.

The astute reader may already have noticed that the grouping or **binning** of the data means that the bar with a midpoint of 10 in Figure 1.1b represents 2 worms scored as 9.5 and one scored as 10, so we have lost some of the information that was visible in Figure 1.1a. Note that we must agree that scores lying on the cut-points always go into the interval above, so the worm whose length is scored as 10.5 is classed as having a length between 10.5 and 11.5.

The two graphs (Figures 1.5a and b) show the same data plotted with different groupings to illustrate the principle of the relative frequency density scale. In Figure 1.5a, we can read off that the probability a worm in the sample is between 5.5 cm and 6.5 cm is $0.1 \times 1.0 = 0.1$. Notice that this depends on the arbitrary fact that we have decided to make the class width exactly 1 cm wide. In Figure 1.5b, the information about this worm is drawn to span the range 5.5–7.5 because we have decided to group the worms into 2-cm classes instead of 1-cm classes. There is still only one worm there though because there were none in the 6.5–7.5 cm class. We now read off from the graph that the probability a worm is in the class 5.5–7.5 is $density \times classwidth = 0.05 \times 2.0 = 0.1$, the same answer as before. To check that you have understood this, show that the probability of a worm in the sample having a length in the range 9.5–11.5 is 0.4, whichever of the two graphs you use.

The advantage of drawing things this way is that we can now quite correctly read off that "the probability a worm in this sample lies between 9.5 and 10.5 cm is 0.3." Let us consider what the histogram for the entire population of worms would look like. We do not know exactly how many there are, so instead of having a scale of frequency of worms up the side of our graph, we could record the *probability* of a worm in the sample falling

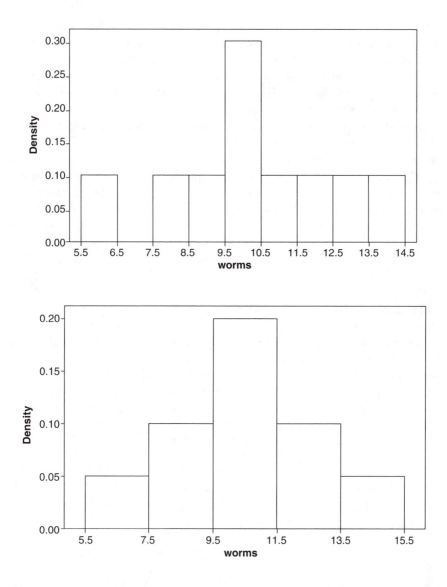

**FIGURE 1.5**
(a) The relative frequency density of the worm sample (see text for explanation). Because the class width is 1 cm this graph also gives the probability that a worm is in a particular 1 cm size class. (b) The relative frequency density plot for worms in 2 cm size classes. Here the probability of a worm lying in any division is class width × density. $0.05 \times 2 = 0.1$ for the leftmost class.

into each class. This would go from 0 to 1. If all worms in the field were 10 cm long, the probability that a worm is in the size class 9.5–10.5 would be 1, which represents certainty that this is the case. However, in reality, the worms vary in length, so the different size classes occur with different probabilities, with the highest being in the central size class. As discussed briefly in Section

1.3, it is a useful fact that measurements like this almost always follow the Normal distribution curve (Figure 1.2); unfortunately, it is much less easy to explain why this is so than it was for the Binomial distribution in Section 1.4. However, what we can say is that the idea of the area under sections of the curve representing probabilities of measurements lying in that region is just the same as that illustrated in Figure 1.4.

Because the equation of the Normal curve is known, a mathematical technique (integration) can be used to work out the area underneath it. The area under the whole Normal curve contains the whole population, so it must be equal to 1 because it is certain (p = 1.0) that every measurement lies somewhere along the x-axis beneath the curve. More helpfully, integration can be used to find out the area between any two length values. This area (strictly, its ratio to the whole area) represents the probability that a member of the population has a length in the specified range. For example, in Figure 1.6, the area between X = 6 cm and X = 7 cm is 0.05. This represents the probability of a worm being between 6 cm and 7 cm long, which is 0.05/1.0 (or only 5% of the population).

Fortunately, we do not need to know anything about either the equation of the curve or about integration to be able to answer such questions. Any information we might need is already available in tables (Appendix C, Table C.1), and most statistically orientated computer programs incorporate them, so that we usually do not need to consult the tables themselves. To use the table we need to calculate the standard deviation (Section 1.2.3; see also Section 1.6.4). The two values where the curve changes direction (points of inflection) lie

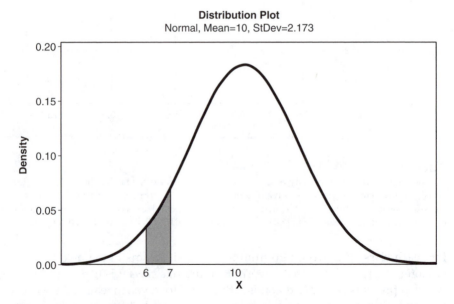

**FIGURE 1.6**
The probability that a member of the population has a length between 6 and 7 cm.

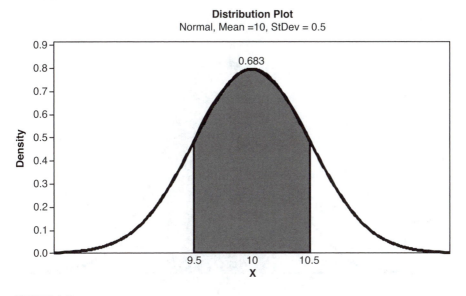

**FIGURE 1.7**
The probability that a member of the population (standard deviation 0.5) has a length between 9.5 and 10.5 cm.

exactly 1 standard deviation below and above the mean (this is not coincidence; it follows from the equation). As we have already seen, the probability of a worm's length being in the range 1 standard deviation either side of the mean is about 0.68 (68%) (Figure 1.7). For example, if the mean is 10 cm and the standard deviation was calculated to be 0.5 cm, we would know that 68% of worms would have lengths between 9.5 cm and 10.5 cm. If the population was more variable and had a standard deviation of 2 cm, we would expect 68% of worms to have lengths between 8 cm and 12 cm (Figure 1.8).

### 1.5.1 Standardising the Normal — An Ingenious Arithmetical Conjuring Trick

The problem with the deductions in the previous section is that they only seem to apply to a particular distributions of worms — those with a mean of 10 and a specified standard deviation. This seems a bit limiting. However, by means of a crafty mathematical trick, we can convert any distribution into a standard form.

- If we subtract the mean from each data point, we create a new scale of measurement in which each number is the distance of the point (its deviation) from the mean, and the mean of the new numbers must necessarily be 0.
- By dividing these deviations by the standard deviation, we convert them into standardised units. We now have numbers representing the number of standard deviations that each point lies above or below the mean.

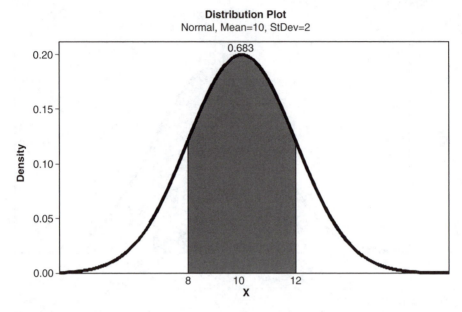

**FIGURE 1.8**
The probability that a member of the population (standard deviation 2.0) has a length between 8 and 12 cm.

This process is known by several different names. **Standardisation,** though commonly used, is perhaps too vague because, in principle, we might apply it to any standard scaling operation. The name **z-transformation** is often used because the result of the change of scale (transformation) is a quantity that is often referred to as a **z-score.** We can describe the process of z-transformation mathematically as

$$zscore = \frac{x - mean(x)}{sd(x)}$$

In Figure 1.9 we show the z-scores calculated for the worm data set. Notice that the shape of the distribution and the relative scatter of the points is unaffected; the only difference is in the scale. However, we now have the huge convenience of being able to read information about the distribution

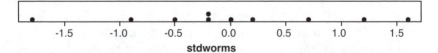

**FIGURE 1.9**
Dot plot of z-scores for worms.

directly from the graph. If the original distribution was reasonably Normal, we can say of the transformed scores (Figure 1.9) that

- The mean of the distribution is 0 because the scale is now deviations about the mean.
- The standard deviation is now 1 because we are expressing it in units of standard deviation.
- Of the total observations, 68% should lie in the range −1 to +1 (approximately).
- Of the total observations, 96% should lie in the range −2 to +2.

These are the properties of the **Standard Normal Distribution**. They will come in extremely useful in the next part of the argument.

### 1.5.2 Estimating Population Parameters from Sample Statistics

Deductions about how likely we are to meet worms of differing lengths can only be made if we know the mean and the standard deviation of the population of worms in which we are interested. Thus, an important part of what follows concerns the relationship between the properties of a sample (sample statistics), which we can measure, and those of the population from which it is drawn (parameters), which we usually cannot measure. It is a useful fact, which we state without proof for the moment, that if a small sample is taken from a large population of continuous measurements, then

- The sample mean is a reasonable estimate of the mean of the population.
- The sample standard deviation is a reasonable estimate of the standard deviation of the population.

It is also an extremely useful fact that

- The means of samples of measurements drawn from a Normally distributed population will themselves be approximately Normally distributed, the more so the bigger the sample.
- More surprisingly, the means of several samples drawn from almost any shaped distribution will be approximately Normally distributed provided the samples are big enough.

As we have already mentioned, measurements such as lengths of worms often follow a Normal distribution, so we can make use of these convenient facts to discover the mean and standard deviation of the population on the basis of the information present in our sample of measurements.

## 1.6   Expressing Variability

Finally, we justify the calculations involved in finding our preferred measure of variability, the standard deviation. We begin by considering from first principles how we might express variability.

### 1.6.1   The First Step — The Sum of the Differences

A first attempt at finding out the variability of our initial sample is to calculate the mean (Section 1.2.2), and then note how far away each value is from it. Now add up the 10 distances or differences (Table 1.4). The sum of the differences would give us a measure of how much scatter there is in the observations.

As we can see in column b, adding up the differences will always give a sum of zero because the mean is defined as the value around which the values in the sample balance. However, as a measure of scatter about the mean, it does not matter whether the deviations are positive, so we discard the sign and add up the absolute deviations — the answer is 16.

If 5 of the worms in our sample were each 9 cm long and the other 5 were 11 cm long, this method would give a mean of 10 cm. Intuitively, we might consider such a sample to be less variable, and indeed the sum of differences in such a case would be 10 (Figure 1.10). So, the less the variability in the observations, the smaller the sum of the differences. This seems a promising way of summarising variability, but we can improve on it.

### 1.6.2   The Second Step — The Sum of the Squared Differences

By our "sum of differences" method, a sample containing just 2 worms, which are 6 cm and 14 cm in length, is no more variable than one that has

**TABLE 1.4**

The Sum of the Differences for the Worms Data

|  | b<br>Difference | c<br>Difference |
|---|---|---|
| Observation (cm) | from Mean Value (cm) | (Positive) (cm) |
| 6.0 | 4.0 | 4.0 |
| 8.0 | 2.0 | 2.0 |
| 9.0 | 1.0 | 1.0 |
| 9.5 | 0.5 | 0.5 |
| 9.5 | 0.5 | 0.5 |
| 10.0 | 0.0 | 0.0 |
| 10.5 | 0.5 | 0.5 |
| 11.5 | 1.5 | 1.5 |
| 12.5 | 2.5 | 2.5 |
| 13.5 | 3.5 | 3.5 |
| Sum | 0.0 | 16.0 |

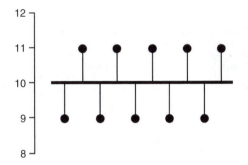

**FIGURE 1.10**
Ten differences of 1 cm.

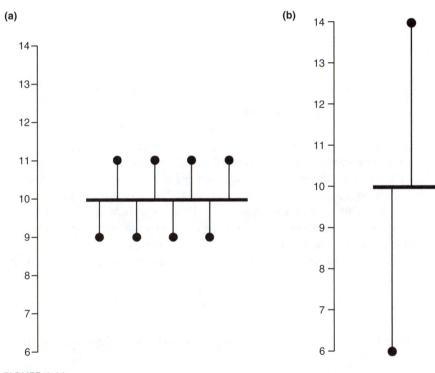

**(a)**

**(b)**

**FIGURE 1.11**
(a) Sum of differences = 8 (consistent). (b) Sum of differences = 8 (variable).

4 worms of 9 cm and 4 worms of 11 cm, because both have a sum of differences of 8 (Figure 1.11). But, from looking at the values, the first sample would seem to be more variable than the second. In the second sample we have many observations that all fall only a small distance from the mean, which is evidence of consistency, whereas in the first sample the results suggest inconsistency; there are only a few values, and these fall far away from the mean. A neat way of taking this into account is to square each difference (to multiply it by itself) before adding the resulting values up (Table 1.5).

**TABLE 1.5**

The Sum of the Squared Differences for Two Other Samples of Worms

| Sample 1 Measured Values (cm) | Difference (All Positive) (cm) | Difference Squared |
|---|---|---|
| 6 | 4 | 16 |
| 14 | 4 | 16 |
| Sum | 8 | 32 |
| **Sample 2 Measured Values (cm)** | | |
| 9 | 1 | 1 |
| 9 | 1 | 1 |
| 9 | 1 | 1 |
| 9 | 1 | 1 |
| 11 | 1 | 1 |
| 11 | 1 | 1 |
| 11 | 1 | 1 |
| 11 | 1 | 1 |
| Sum | 8 | 8 |

The sum of the squared differences between the observations and their mean is 8 for the consistent sample (sample 2) but 32 for the inconsistent one (sample 1); so this revised method gives a much better picture of the variability present than just the sum of the differences suggested in Section 1.6.1. The phrase "the sum of the squared differences between the observations and their mean" is usually abbreviated to "the **sum-of-squares**." Unfortunately, this can lead to a misunderstanding about how it is calculated. It does *not* mean the sum of the squares of each observation, as the following example should make clear.

The sum-of-squares for a sample of three worms measuring 2, 3, and 4 cm, respectively, is *not*

$$(2 \times 2) + (3 \times 3) + (4 \times 4) = 29$$

Instead, because the mean of the values 2, 3, and 4 is 3, the sum-of-squares is

$$(2 - 3)^2 + (3 - 3)^2 + (4 - 3)^2 = 1^2 + 0^2 + 1^2 = 2$$

A better name would be **sum of squared deviations about the mean**.

### 1.6.3   Third Step — The Variance

So far so good, but we have ignored the number of worms in the sample in our calculation. Each worm measured counts as an **observation**. The more worms we measure, the greater will be the sum of squared deviations about the mean *simply because we are adding up more squared differences*. To take account of this fact and to provide a standard estimate of variability, we divide the sum of squares by the number of observations minus one. Generally,

$$\text{Sample Variance} = \frac{\sum_{i=1}^{n}(x_i - \bar{x})^2}{(n-1)}$$

In this case, we divide the sum of squares by nine if we have ten worms or by seven if we have eight worms.

### 1.6.3.1 Degrees of Freedom

Why do we divide by one fewer than the number of observations rather than the number of observations itself? The justification for this is the *number of independent pieces of information about the variability in the population*. If we have a sample of only one worm and it is 9 cm long, we have an estimate of the population mean (9 cm) but we have no information about how variable is the population. As soon as we select a second worm of length 10 cm, we have two values (9 cm and 10 cm). We can revise our estimate of the population mean to (9 + 10)/2 = 9.5 cm, and we now have one piece of information about its variability. The difference between the mean and the length of the first worm is 9.0 – 9.5 = –0.5. What about the other worm? Well, the deviation is (10 – 9.5) = +0.5, but if we knew the first one, we must already know the second because they sum to 0 *by definition* (see column b in Table 1.4), so there is only one independent item of information about the variability. With three worms we can revise our estimate of the mean again, and we now have two *independent* pieces of information:

Worm lengths 9.0, 10.0, 11.0 New mean = 10.0

Two independent pieces of information about variability are: (9.0 – 10.0) = –1 and (10.0 – 10.0) = 0. The third difference (11.0 – 10.0) is not independent because, given the first two, it must have the value 1.0 if the deviations are to sum to zero.

It is common to refer generally to the number of observations in a sample as *n*. So, in our main example, where *n* = 10, we divide the sum of squares of our sample by *n* – 1 (in our case 9) to get the **sample variance.** The number *n* – 1 is referred to as the **degrees of freedom** because it refers to the number of independent items of information about variability in the sample.

### 1.6.3.2 Estimated Parameters

Another way of looking at the problem of degrees of freedom is to see that the three differences (9.0 – **10.0**), (10.0 – **10.0**), and (11.0 – **10.0**) all contain the sample mean (**10.0**), which was *estimated* using the other 3 numbers. Thus, you could say that there *are* three independent bits of information here but that you could represent them either as the three observations or as the calculated mean and any two of the observations. Because the variance is to be calculated as deviations about the estimated mean, there can only be $n - 1 = 2$ degrees of freedom in this case. This leads us to a more general formulation that the degrees of freedom associated with any measurement of variation based on a sample of observations are

*number of observations − number of estimated parameters*

In this case we estimate just one parameter, the mean, but later in the book we will meet situations where more than one parameter is being estimated and degrees of freedom are not simply $n - 1$.

### 1.6.3.3   Bias in the Estimation

Finally, there is a practical justification for using $n - 1$ to calculate the sample variance. Statisticians have shown (using many lines of algebra — see, e.g., Grafen and Hails (2002)) that a sample variance calculated by dividing the sum of squares by $n - 1$ is an **unbiased estimate** of the population variance. This means that if we took many samples and calculated their variances by dividing by $n - 1$, their average would be closer to the population variance than it would be if we divided by $n$; it would be neither consistently too big nor consistently too small. If you do the obvious thing of dividing by $n$, the estimate is **biased**. It would be consistently too small. We will come to why this is undesirable later.

---

## Box 1.4 Formula for the Population Variance

In the unlikely event of your having an entire population to play with, the population variance *would* be the sum of squared deviations about the population mean $\mu$ divided by N, where N is the total number of items in the population:

$$\text{Population Variance } \sigma^2 = \frac{\sum_{i=1}^{N}\left(X_i - \mu\right)^2}{N}$$

If the entire population consisted of just one worm, the variation in the population actually would be zero. The population variance is *defined* as the squared deviations of the individual measurements about the true population mean, so the issue of estimation does not arise, and N is the correct denominator. The population standard deviation would just be the square root of the population variance, in the usual way.

Many statistical calculators have a button for the sample standard deviation and another for the population standard deviation. It is very important to be sure which one you should be using; nearly always it will be the sample standard deviation that is relevant.

### 1.6.4   Fourth Step — The Standard Deviation

We have now obtained variances (measures of variability within a sample) that can be compared with one another, but they are in units that are different from those used in the initial measurement. If you go back to Section 1.6.2, you will see that we took the differences in length (cm) and squared them. Thus, in our example, the units of variance are in $cm^2$. We would not naturally think of the variability of the length of a worm in $cm^2$, so we take the square root of the variance to return to cm. The result is called the **standard deviation** and is the quantity referred to in Section 1.2.3 and again in Section 1.4 and Section 1.5.

We can now work out the standard deviation from our original data set.

The sum of squares is 42.5 (Table 1.6). So, the variance is 42.5/9 = 4.722, and the standard deviation, which is the square root of this, is 2.173. This value helps us to judge the extent to which worm length varies from one individual to another, but its usefulness becomes clearer when we put this piece of information together with our knowledge of the Normal distribution (Figure 1.7) Remember that this told us that 68% of the worms in the field will have lengths between

$$mean \pm 1 \times s.d.$$

**TABLE 1.6**

The Sum-of-Squares for the Worms Data

| Observation | Difference (All Positive) | Difference Squared |
|---|---|---|
| 6.0 | 4.0 | 16.0 |
| 8.0 | 2.0 | 4.0 |
| 9.0 | 1.0 | 1.0 |
| 9.5 | 0.5 | 0.25 |
| 9.5 | 0.5 | 0.25 |
| 10.0 | 0.0 | 0.0 |
| 10.5 | 0.5 | 0.25 |
| 11.5 | 1.5 | 2.25 |
| 12.5 | 2.5 | 6.25 |
| 13.5 | 3.5 | 12.25 |
| Sum | 16.0 | 42.5 |

In our case that is between

$$10 \text{ cm} - 2.173 \text{ cm and } 10 \text{ cm} + 2.173 \text{ cm}$$

which works out to be between

7.827 cm and 12.173 cm, often written as [7.827, 12.173] cm

This sounds amazingly precise — it implies that we can measure the length of a worm to the nearest 0.01 of a millimetre! Because we only measured our worms to the nearest 0.5 cm (5 mm), it is better to express this result as

Of the worms in the field, 68% have lengths between 7.8 cm and 12.2 cm.

If the worms we measured still had a mean length of 10 cm but were much less variable — for example, with lengths mainly in the range from 9 to 11 cm — the standard deviation would be much less. Try working it out for these values:

8.5, 9.0, 9.5, 9.5, 10.0, 10.0, 10.5, 10.5, 11.0, 11.5

(you should get 0.913).

Remember, though, that these statements about all the worms in the field depend on two claims that we have yet to justify:

- That the sample mean is a reasonable estimate of the population mean
- That the sample standard deviation is a reasonable estimate of the population standard deviation

### 1.6.5   Fifth Step — The Sampling Distribution of the Mean

Measuring ten worms gives us an estimate of their mean length and of their variability, but we have already implied that, if we took a second sample of worms, the estimate of the mean would be slightly different. Thus, in this section we look at how the mean varies from one sample to another. This is referred to as a **sampling distribution of the mean**.

As an illustration, imagine that the whole population consists of only six worms and we are going to estimate the mean length of worms in the population by sampling just two of them (the two worms being returned to the population after each sampling). Table 1.7 is a list of all possible outcomes.

For convenience, where the two worms in a sample differ in length, the shorter has been called worm 1. There are 15 possible samples (A to O)

**TABLE 1.7**

The 15 Possible Samples of 2 Worms from a Population of 6 Worms

| Sample | A | B | C | D | E | F | G | H | I | J | K | L | M | N | O |
|--------|---|---|---|---|---|---|---|---|---|---|---|---|---|---|---|
| Worm 1 | 8 | 8 | 8 | 8 | 8 | 9 | 9 | 9 | 9 | 10 | 10 | 10 | 10 | 10 | 11 |
| Worm 2 | 9 | 10 | 10 | 11 | 12 | 10 | 10 | 11 | 12 | 10 | 11 | 12 | 11 | 12 | 12 |

Lengths 8 9 10 10 11 12: Population mean 10 cm

**TABLE 1.8**

The Results of Taking All Possible Samples of Two Worms from a Population of Six Worms

| A<br>Sample D | B<br>Number of Samples | C<br>Sample Mean | D<br>B × C |
|---------------|------------------------|------------------|------------|
| A | 1 | 8.5 | 8.5 |
| B,C | 2 | 9.0 | 18.0 |
| D,F,G | 3 | 9.5 | 28.5 |
| E,H,J | 3 | 10.0 | 30.0 |
| I,K,M | 3 | 10.5 | 31.5 |
| L,N | 2 | 11.0 | 22.0 |
| O | 1 | 11.5 | 1.5 |
|   | Total = 15 |   | Total = 150<br>Mean = 150/15 = 10.0 |

because that is the maximum number of different combinations of pairs of worms that you can get from 6 worms. Because two of the worms share the same length (10 cm), some samples will give the same mean (see samples B and C, for example).

We can see that, by chance, our sample might have provided us with an estimate of the mean that was rather extreme: sample A would give 8.5, whereas sample O would give 11.5, compared with a population mean of 10.0. The means of the 15 possible samples can be summarised as in Table 1.8.

The second column in this table shows that, if we take one sample of 2 worms, we have a 1 in 15 chance of selecting sample number A and so getting an estimated mean of 8.5. However, we have a 3 in 15 (or 1 in 5) chance of getting a sample with a mean of 9.5 (samples D, F, or G) or of 10.0 (E, H, or J) or of 10.5 (I, K, or M). This is the **sampling distribution** of the sample mean. It shows us that we are more likely to obtain a sample with a mean close to the population mean than we are to obtain one with a mean far away from the population mean.

Also, the distribution is symmetrical (Figure 1.12) and follows the Normal curve, showing that, if we take a series of such samples, we will get an unbiased estimate of the population mean.

The mean of all 15 sample means equals the population mean (10.0, bottom right of Table 1.8), which is the result we set out to prove in this section.

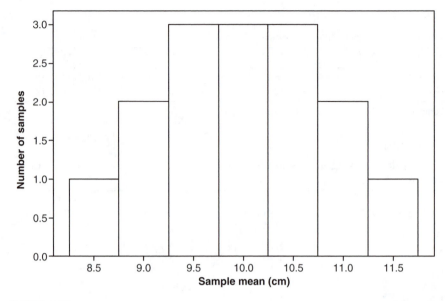

**FIGURE 1.12**
The sample means of a large number of samples from the same population.

### 1.6.6   Sixth Step — The Standard Error of the Mean

In reality, we will only have one sample mean, and we will not know the true population mean because that is one of the things we are trying to estimate. Therefore, we need a way of expressing the variability from one sample mean to another. For example, it is very common in experiments (Chapter 4) to want to compare, say, the mean length of 10 worms fed on one type of food with the mean length of 10 worms fed on a different type of food.

To estimate the reliability of the mean, a slightly different statistic is used: the **standard error** (s.e.). In contrast to the standard deviation, the standard error describes our uncertainty about how well the mean of a sample of a given size estimates the population mean of the population from which the sample was drawn. In Section 1.6.5 we said that the sampling distribution of the mean is itself Normal. In other words, if we took a large number of samples from the same population and plotted the means out as a frequency distribution, it would look like (and actually be) a Normal distribution. The mean of the sample means would be equal to the population mean. We have demonstrated this on a very small scale in Figure 1.12. You can also see that an obvious measure of the reliability of a sample mean as an estimate of the population mean would be the standard deviation of the sampling distribution (i.e., the standard deviation of Figure 1.12). It turns out that the standard deviation of the sampling distribution is related to the population standard deviation $\sigma$ by the following formula:

$$sd(sampling\ distribution) = \frac{\sigma}{\sqrt{n}}$$

provided we are dealing with a small sample from a large population. (Note that this is not in fact the case for Figure 1.12 because a sample of 2 from a population of 6 is not a small sample from a large population — an additional trick is needed in this situation, beyond the scope of this book).

We call the standard deviation of the sampling distribution the **standard error of the mean**, or **standard error** for short.

Notice that this can be expressed in three equivalent ways:

$$Standard\ error = \sqrt{\frac{variance}{sample\ size}} = \frac{\sqrt{variance}}{\sqrt{sample\ size}} = \frac{standard\ deviation}{\sqrt{sample\ size}}$$

The standard error gets smaller as the sample size increases, because the standard deviation is divided by the square root of the sample size. So, if our sample size was 16, the standard error of the mean would be the population standard deviation divided by four. If we took a sample of 25, we would divide the standard deviation by five, and so on. Intuitively, this makes sense because more information (bigger $n$) should lead to greater precision in the estimate (smaller s.e.). The smaller the variance, the smaller the standard deviation, and this also makes sense because samples drawn from a less variable population should be more consistent than those drawn from a highly variable population.

There remains a problem, which is that the standard error as defined in the preceding text involves the *population* standard deviation, $\sigma$, which we usually do not know. In practice, we have to rely on the claim that the sample standard deviation, $s$, is a reliable estimate of the population standard deviation, $\sigma$. This means that we can use the quantity

$$s.e. = \frac{s}{\sqrt{n}}$$

in circumstances when the estimation of $\sigma$ by $s$ can be considered reasonable.

You should take a moment to contemplate what an incredibly powerful tool Figure 1.12 represents. Because we know that the sampling distribution is Normal, and we know its two defining parameters (its mean and its standard deviation), we know everything about it. Thus, if the population mean is 10, then the mean of a sample of $n = 2$ is not very likely to be as extreme as 8.5 or as big as 11.5, as we saw when we enumerated all the possibilities (Table 1.8). Of course, in the real world we cannot enumerate all possible samples, so we must rely on the theory. Back in Section 1.6.4 the standard deviation of the sample of ten worms was found to be 2.173 cm.

The standard error of the mean will be less because we divide the standard deviation by the square root of 10 (which is 3.162).

s.e. = 2.173/3.162 = 0.69 (we should perhaps round this to 0.7 cm)

So, the sampling distribution for the means, of which this particular sample mean is one member, is a Normal distribution with mean 10 and standard deviation 0.7. This is a *hypothetical* distribution of the sample means we *would* get if we drew a large number of samples of $n = 10$ from a population whose true mean is 10.0 and whose true standard deviation is 2.173.

Exactly what we can do with this information is explored further in Chapter 2.

# 2

# Confidence Intervals

> If you want to inspire confidence, give plenty of statistics. It does not matter that they should be accurate, or even intelligible, as long as there is enough of them.
>
> —Lewis Carroll

## 2.1   The Importance of Confidence Intervals

Any estimate that we make of the value of a population parameter (such as the mean length) should be accompanied by an estimate of its variability — the **standard error**. As we have seen in Chapter 1, we can say that there is a 68% chance that the population mean will lie within one standard error of the sample mean of any sample that has been taken correctly. This range is called a **68% confidence interval**. However, we usually want to be rather more confident than only 68%. The greater the chance of our range containing the population mean, the wider the confidence interval must be.

Of course, there is a limit to this: if we want to be absolutely certain that the range contains the population mean, we need a 100% confidence interval, and this must be very wide indeed; it must contain the whole of the sampling distribution curve (Figure 1.7), not just the fat middle part. We would then be certain that the true value lay within the confidence interval, but the interval would be so wide that it would give us little useful information about what the value actually was. A good compromise between certainty and precision is usually taken to be a 95% confidence interval. We can think of the 68% confidence interval as being obtained by multiplying the standard error by one; therefore, to calculate the 95% confidence interval, we need a "multiplier" of greater than one. Statistical tables, summarising the Normal distribution curve, tell us that we need to multiply by 1.96 for a 95% confidence interval (see Appendix C, Table C.1, where the multiplier = $z$ = 1.96 for $p$ = 0.975, i.e., 97.5% of the area lies below this point, so 2.5% lies above it and because the curve is symmetrical, 2.5% lie below −1.96; this detail is discussed further in Section 2.5). More generally, we can write:

$$confidence\ interval = mean \pm (z \times SE\ mean).$$

Unfortunately, there is one further catch. The conclusions about the Normal distribution only apply if the sample size is reasonably large (about 30 or more). To deal with small samples such as the worms in Chapter 1, we have to take additional measures, which we will come to later (Section 2.7). Meanwhile, we introduce another example.

## 2.2   Calculating Confidence Intervals

Imagine that you are the owner of a fishing boat and you are wondering whether to start catching and bringing to market a particular species of fish. Market research indicates that there is a demand for these fish, but it is only economical if we can land 40 t or more in an average week of fishing. If the mean weight of your catch was less than this, then the cost of putting to sea to catch the fish would be greater than the likely return, so you would probably decide to go after some other species.

The way to find out whether we could expect the mean of our catch to be more than 40 t is to take a sample (Chapter 4). We would deploy our boats for a series of week-long trials to see what the mean tonnage actually was for each week. An example of such a set of data is found in Table 2.1, which shows the mean tonnage caught in 30 weeks of fishing for Patagonian toothfish (*Dissostichus eleginoides*).

The results from the sample catches show that, on average, the mean weight of each catch is just over 41 t. If every possible catch that could be taken from the Southern Ocean (i.e., the population of catches) had been made, we would know that 41.37 t was the true mean value. However, as we have only *sampled* some of the population of possible catches, we have only an *estimate* of the mean. On the basis of the evidence we have, the best guess is that the mean is 41.37 t, but we know that there is some uncertainty about how close the true population mean is to this estimate. As you can see (Figure 2.1), some catches were as small as 37 t, whereas for others it was as much as 45 t. The standard deviation of our sample is, in fact, 2.68 t, and we had 30 catches in the sample.

**TABLE 2.1**

Mean Weight of 30 Catches of Patagonian Toothfish (Tonnes)

| Data Display | | | | | | | | | | | | | | |
|---|---|---|---|---|---|---|---|---|---|---|---|---|---|---|
| *Toothfish* | | | | | | | | | | | | | | |
| 44 | 43 | 35 | 39 | 41 | 45 | 36 | 40 | 40 | 45 | 42 | 43 | 38 | 44 | 44 |
| 40 | 43 | 39 | 42 | 43 | 42 | 44 | 39 | 44 | 42 | 42 | 42 | 44 | 39 | 37 |

To repeat, we know from the discussion in Chapter 1 that this sample mean is more likely to be close to the true mean than it is to be far away from it. In Section 2.1 we showed that, in theory, it will be within 1.96 s.e. of the mean 95% of the time because of the properties of the normal distribution. Thus, for our sample, we find that the true mean is likely to lie in the range

$$\bar{x} \pm 1.96 \times \frac{\sigma}{\sqrt{n}}.$$

Because 30 is a reasonably large sample, we can assume that the sample standard deviation is a reasonable estimate of the population standard deviation, and hence, we may say that the true mean lies in the range

$$41.37 \pm 1.96 \times \frac{2.68}{\sqrt{30}} = 41.37 \pm 1.96 \times 0.49$$

which is about 40.36 to 43.36 t (Figure 2.1). So, given these data, we are 95% confident that the population mean lies in the range 40.41 to 42.33.

Another way of representing this can be seen in Figure 2.2, which shows the sampling distribution for the mean based on the evidence of this sample. The most likely value for the true mean is 41.37, the sample mean, and the curve shows the probability density (see Chapter 1).

If we have worked out that, to make fishing worthwhile, the mean catch needs on average to be 40 t or more, we might decide that we are happy to proceed because there is only a small chance (2.5%, which is the area in the

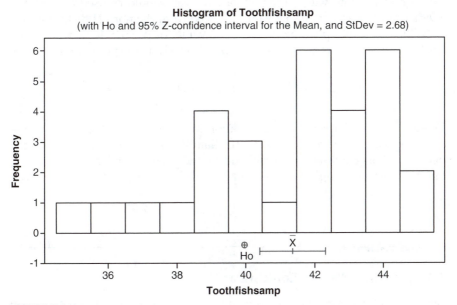

**FIGURE 2.1**
Histogram of mean weight of 30 catches of Patagonian toothfish (tonnes).

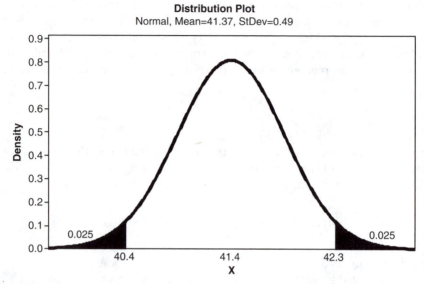

**FIGURE 2.2**
Sampling distribution for the mean catch tonnage based on the evidence from 30 catches of Patagonian toothfish (tonnes).

left-hand tail of the distribution shown in Figure 2.2) that the population mean will be less than 40.4 t.

However, we may decide that a 2.5% chance of bankruptcy from this source of risk is too much and that we are only prepared to take a 0.5% risk. Then, a 99% confidence interval for the population mean should be used instead. For a 99% confidence interval, the value we need from the normal distribution is 2.58 (Table C.1, interpolating between the value for p = 0.9946 (2.55) and p = 0.9953 (2.60)), so the range that is 99% likely to include the population mean is

$$41.36 \pm 2.58 \times 2.68 \Big/ \sqrt{30} = 41.36 \pm 1.262 = (40.10, 42.62)$$

There is a 0.5% chance of the population mean being less than 40.10 t (the 0.5% chance of the mean being more than 42.62 t does not influence the decision). This confidence interval still excludes the threshold value of 40 t, but only just. Although we could argue that the fishery will be economical, it is a close run thing.

## 2.3  Another Way of Looking at It

We could reformulate the question like this. Given our single sample of 30 catches, which has mean 41.36 t and standard deviation 2.68 t, how likely is it that the true mean is as much as $41.36 - 40 = 1.36$ t below the sample mean? In other words, we look at the distribution of the *difference* we expect between the sample mean and the true mean, $\bar{x} - \mu$. In Chapter 1 we showed that the average of a series of sample means $\bar{x}$ drawn from a population whose true mean is $\mu$ would be

$$mean(\bar{x}) = \mu$$

By $mean(\bar{x})$ we mean "the mean of several sample means" or the mean of the sampling distribution for $\bar{x}$. So, it follows that

$$mean(\bar{x}) - \mu = 0.$$

We also saw in Chapter 1 that the standard deviation of the sampling distribution for the mean is

$$\sigma / \sqrt{n},$$

which we will estimate as

$$s / \sqrt{n} = 2.68 / \sqrt{30} = 0.49.$$

So we can say that the distribution of $\bar{x} - \mu$ for our 30 sample means is itself Normal, with mean 0 and standard deviation 0.49. However, the standard deviation is not changed by subtracting the mean from all the observations, so the standard deviation of $mean(\bar{x}) - \mu$ is also

$$\sigma / \sqrt{n}.$$

We can therefore use the same rescaling trick we met in Section 1.5.1. We divide the difference between the sample mean and the population mean by its standard deviation

$$(\text{in this case } \sigma / \sqrt{n})$$

to convert it into a standardised normal distribution

$$z = \frac{(\bar{x} - \mu)}{\sigma / \sqrt{n}}.$$

Recall (Section 1.5.1) that this is normal with mean 0 and standard deviation 1. We can now reformulate our question about the fishery as follows: If the true mean catch was only 40 t, how likely is it that we managed to get a sample mean of 41.37? We plug the numbers from sample catches into the equation we calculate:

$$z = \frac{\left(\bar{x} - \mu\right)}{\sigma/\sqrt{n}} = \frac{\left(41.367 - 40\right)}{2.68/\sqrt{30}} = 2.79$$

and the question now becomes, "How likely is a value as large as 2.79 to turn up in a Normal distribution whose mean is 0 and whose standard deviation is 1?" Because we already know that in any Normal distribution, 95% of values lie within 1.96 standard deviations either side of the mean, we can say that 2.79 lies well outside this region. Consequently, it is rather unlikely that the true mean is as low as 40 t (Figure 2.1 and Figure 2.2).

It is very important to see that what has just been done here is simply a rearrangement of the information used to compute confidence intervals in Section 2.2.

## 2.4    Your First Statistical Test

This section uses a formalization of what we have been describing in words to give us a **statistical hypothesis test**. Here we want to test whether the mean weight of a catch is 40 t or less. We work through this example in two ways to show that the following are very closely related and rely on the same idea; that we can estimate properties of a population from a single sample:

- *Estimation* (estimating the true mean using a sample)
- *Hypothesis testing* (statistically testing whether the sample or true mean is the same as the value we are interested in)

The procedure followed in conducting a hypothesis test is completely general, and its logic needs to be thoroughly understood. A bit of extra effort here will pay handsome dividends later on.

### 2.4.1    Develop a Research Hypothesis — Something That You Can Test

The first step is to decide what we are trying to find out. In this case, our **research hypothesis** is that it might be worthwhile starting to catch toothfish because, on average, the catch we can make per week is more than 40 t. Notice that, at this stage of the procedure, we have not yet collected any data, so we do not have a clue what the actual mean tonnage will turn out to be.

### 2.4.2 Deduce a Null Hypothesis — Something That You Can Disprove

The idea behind this is that you can never prove a generalisation (e.g., "All swans are white") beyond doubt, but you can disprove it (by finding a nonwhite one). This is a very simplified version of what philosophers of science call the **problem of induction**, which is essentially the puzzle of our ability to generalise from our experience, the basis of scientific discovery. In principle, you might think that you would be more confident about the whiteness of swans after you had observed 100,000 of them than if you had only observed 100. This would be correct if the world contained a finite number of swans (say, for the sake of argument, 1 million). Your increased confidence would be related to the fact that 100,000/1,000,000 = 1/10 is a much bigger number than 100/1,000,000 = 1/10,000. Another way to view it is that, after seeing 100,000 of them, there are fewer left still to see than there were after the first 100. In most circumstances, however, we have no idea how many objects the "All" refers to, so we must treat it as infinite. If we apply formal logic to the problem, we reach the conclusion that no matter how many white swans we see, there is still an infinite number we have yet to see, so we are no nearer proving that the generalisation "All swans are white" is true. On the other hand, we can prove that it is false rather easily, simply by producing a nonwhite one (there are quite a few of these, an Australian species, breeding wild in Britain at the moment).

In logic-speak, the Null hypothesis is the **negation** of the research hypothesis. This means that if the research hypothesis is true, then the Null hypothesis is false, and if the Null hypothesis is false, then the research hypothesis is true. Thus, the Null hypothesis here is that toothfish catches on average weigh less than 40 t. Using only samples from an effectively infinite population, we cannot prove conclusively that this is true. However, we would be able to prove that it is false, or at least very unlikely to be true, by showing that our catches are nearly always heavier than that. As we have already said, if the Null hypothesis *is* false, then the research hypothesis *must be true*.

### 2.4.3 Collect the Data

At this stage we work out how to collect some relevant data. We must do this in such a way as to ensure that the test is fair; we will look at this further in Chapter 4. In our example we sent our boats out on a week-long cruise on 30 different occasions and measured the mean fish tonnage at the end of the week. We need to be sure that these are a representative set of samples. For example, we would want to spread them out through the fishing season and over the area of the proposed fishery. We would also need to be certain that a large catch taken in one place did not affect any of the other catches, as they might do if made nearby at the same time.

So, the population is of possible catching expeditions, and the parameter is the mean tonnage yielded each time the boat returns from a week at sea.

### 2.4.4   Calculate a Test Statistic

This is the procedure followed in Section 2.3. We calculate the z statistic as already shown:

$$z = \frac{\left(\bar{x} - \mu\right)}{\sigma/\sqrt{n}} = \frac{\left(41.367 - 40\right)}{2.68/\sqrt{30}} = 2.79$$

Notice again that we have used the sample standard deviation as an estimate of the population standard deviation $\sigma$, which we could not know. Because the sample size is 30, this is reasonable thing to do.

### 2.4.5   Find the Probability of the Test Statistic Value if the Null Hypothesis Was True

#### 2.4.5.1   *Finding the Probability Directly*

If we picked numbers at random from a Normal distribution with *mean* = 0 and *sd* = 1, we would be very unlikely to get one as big as 2.79. Just how unlikely we find by looking at the distribution of the standardised Normal distribution (Table C.1). We look for 2.79 in the column labelled z, and we find next to (the nearest we can get, 2.80) the value 0.9974. This is the probability that we would get a value of 2.80 *or less* in a standard Normal distribution. We show this graphically in Figure 2.3, where the shaded area corresponds to the probability of z ≤ 2.8.

However, we are interested in the probability of getting a number as extreme as 2.80. This actually means as big as 2.80 *or bigger,* which is the part of the distribution to the right of z = 2.80, known as the "right tail" of the distribution. It is certain (p = 1.0) that either z ≤ 2.8 **or** z > 2.80, so we can say

$$p(z > 2.8) = 1 - p(z \leq 2.8)$$

and the probability we want is

$$1 - 0.9974 = 0.0026 \text{ (Figure 2.4)}.$$

#### 2.4.5.2   *Comparing the Test-Statistic with a Threshold*

Another approach is to establish the magnitude of z such that it is exceeded only 5% of the time. We call this the 5% critical value of z. It is often written $z_{crit}$. We used this idea in establishing a 95% confidence interval in Section 2.2; however, pay attention because there is a catch here.

Going back to the tables of z (Table C.1), can we find the value such that the right-tail area is 5% (0.05)? As the p-values in the table are the left-tail probabilities (as shown in Figure 2.3, for example) we must first calculate $1 - 0.05 = 0.95$ to find the probability to look for in Table C.1. The correct

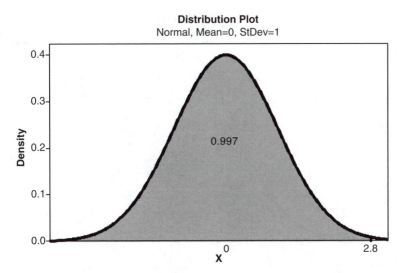

**FIGURE 2.3**
Left-tail probability of Normal distribution. The area to the left of 2.8 is 0.997.

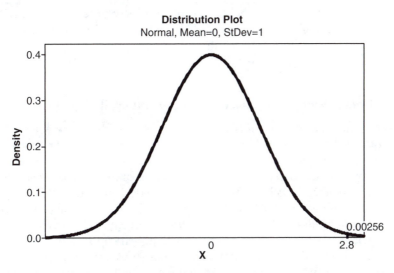

**FIGURE 2.4**
Right-tail probability of Normal distribution. The area to the right of 2.8 is 0.00256.

z-value is approximately 1.64 (Figure 2.5), which is not actually in Appendix C, Table C.1. The closest you can get is 1.65, for which the left-tail probability is 0.9505, close but a bit too big. If we could just remember that the **critical value** of $z$ for which we would reject $H_0$ is 1.64, we can save ourselves the trouble of looking things up in Appendix C, Table C.1.

Thus, if we said we would accept the Null hypothesis only if $z < 1.64$, then we would only reject it incorrectly 5% of the time when really it was true. It is very important to realise that we *would* be wrong about it 5% of the time!

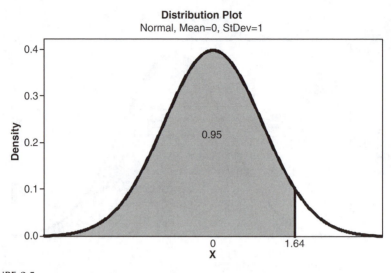

**FIGURE 2.5**
Five percent threshold (one-tailed) of Normal distribution.

Depending on the method chosen, we may now say either that $p(z \geq 2.79) = 0.005$, or alternatively, that because $2.79 > 1.64$, $p(z \geq 2.79) > 0.05$. The probability involved in these statements (whichever you choose to use) is the p-value established by our test.

### 2.4.6 Decide Whether to Reject or Accept the Null Hypothesis

Statistics is not magic; rather, it is a kind of gambling with Nature. We might be unlucky and draw a sample with mean = 41.37 from a population whose true mean is 40 or less and whose standard deviation is 2.68, but this would not happen very often. When it did happen, we would draw the wrong conclusion. This type of mishap is called a **Type I error.** The decision criterion is the probability we are prepared to accept that we would draw the wrong conclusion *in the case that the Null hypothesis is actually true.* Conventionally, statisticians say that they are prepared to accept a risk of 5% that their conclusion is incorrect in this situation, even though they have done their sums correctly.

The precise meaning of the p-value that we use in this decision (calculated in Section 2.4.5) is rather subtle and often misunderstood. If we get a small p-value, then one or other of the following statements can be made:

- The Null hypothesis is false.
- The Null hypothesis is true, but an unlikely event, one whose probability is the p-value, has occurred.

Under these circumstances, it seems more rational to believe that the Null hypothesis is false, than that it is true *and* that something unlikely has happened. This is the basis of the convention, which we will follow, of rejecting the Null hypothesis when the p-value is less than some agreed criterion. For the purposes of the example we will adopt a criterion of 5% or p = 0.05 for the threshold of rejection. We often refer to the criterion probability as α or the α-**level**. So, you may see the form of words "the Null hypothesis was rejected at α = 0.05."

Because in this case z = 2.79, which is considerably more than 1.64, we reject the Null hypothesis because, whichever way we look at it, the p-value is less than our criterion of 0.05 or 5%. The conclusion is that the mean weight of toothfish catches probably is > 40 t.

### 2.4.7 Using the Computer to Conduct the Test

Although the manual calculations for the one-sample z-test are comparatively trivial, we illustrate the use of the computer here because, as things become more complicated, it becomes less and less practical to use manual methods. In MINITAB, the command is:

```
Stat>Basic Statistics>One Sample z-test Summarised data;
Mean 41.37, Sample size 30, Standard deviation 2.68
Options Confidence level: 95 Alternative: greater than
```

Note that we must specify that the "alternative" hypothesis (MINITAB's name for the research hypothesis) is "greater than," because our research hypothesis is that the mean catch weight is *greater than* 40 t. This point is taken up in Section 2.5.

Here is the output:

```
One-Sample Z
Test of mu = 40 vs > 40
The assumed standard deviation = 2.68

                           95%
                          Lower
  N      Mean   SE Mean   Bound     Z      P
 30   41.3700   0.4893   40.5652   2.80   0.003
```

You can see all the quantities we calculated by hand here (ignore the figure called "95% lower bound," which is explained in more detail in the next section).

If you put in the actual sample data to a column called "toothfishsamp," the command sequence is slightly different:

```
Stat>Basic Statistics>One Sample z-test Samples in columns:
toothfishsamp, Standard deviation 2.68 Options Confidence level:
95 Alternative: greater than Graphs: histogram of data
```

This saves us the bother of calculating the sample mean and standard deviation for ourselves, but otherwise gives the same printout (give or take some rounding errors). Entering the data completely in this way also allows

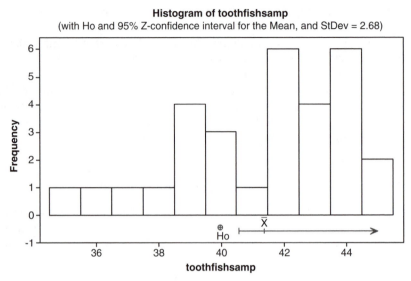

**FIGURE 2.6**
Histogram of mean weight of 30 catches of Patagonian toothfish (tonnes). Confidence interval
with no upper bound.

us to plot Figure 2.6, which shows the whole story. Because we chose to test
the alternative hypothesis that toothfish catches have a mean weight of
*greater than* 40 t, the confidence interval is drawn with no upper bound. As
we saw from the manual calculations, at a rejection criterion of 5%, the Null
hypothesis value of 40 (or less) lies outside the confidence interval, as we
concluded from the manual calculations.

## 2.5   One- and Two-Tailed Tests

The test we conducted in Section 2.4 concerning the mean weights of toothfish captures is called a "one-tailed one-sample z-test" because, in deciding
whether or not to reject the Null hypothesis, we looked only at the right tail
of the z-distribution (Figure 2.5) This was because the research hypothesis
was that the mean capture tonnage was *greater than* 40 t, implying that the
Null hypothesis is that capture tonnages are *less than or equal to* 40 t. Consider
an alternative scenario in which we extend our fishing to a new area of the
Southern Ocean. We are interested in a slightly different question, namely,
whether the catches we can make in the new area are the same as the those
in the area we already know about. For the sake of argument we will say
that, on the basis of a very large number of catches, we know that the mean
catch weight is, in fact, 40 t. The fish in the new area might be either easier
or harder to catch, on average; they might be a different subspecies or variety,
or the growing conditions could be different. So we would have:

- Research hypothesis: Catches in the new area *are different* in mean tonnage from the 40-t standard established in the old area.

- Null hypothesis: Catches in the new area *are not different* in mean weight from 40 t.

We collect our sample data, and (just to keep the sums simple), we find that the mean of 30 catches is 41.367 t with standard deviation 2.68, the same figures as we used before.

The calculation procedure is no different from before — we compute

$$z = \frac{(\bar{x} - \mu)}{\sigma/\sqrt{n}} = \frac{(41.367 - 40)}{2.68/\sqrt{30}} = 2.79 \ .$$

However, because we did not specify in the research hypothesis whether the mean would be more or less than 40 t, we would now want to reject the Null hypothesis if the discrepancy between observed mean and Null hypothesis mean (40) was *either* positive *or* negative. This means that we have to consider the probability that, if the Null hypothesis were really true, we would get a z-score as big as +2.79 or as small as –2.79. Looking at Figure 2.7, we can see that this is the probability to the right of +2.79 *plus* the probability to the left of –2.79, which is

$$0.00264 + 0.00264 = 0.00528$$

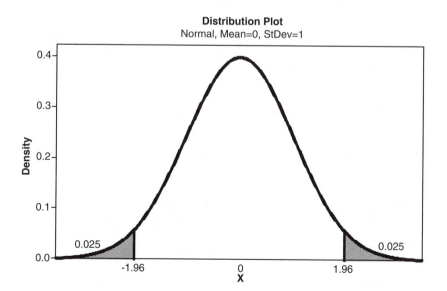

**FIGURE 2.7**
Five percent threshold (two-tailed) of normal distribution.

Because both tails of the distribution of z are involved, we call this a "two-tailed test." To conduct the two-tailed test using MINITAB, all we do is change the alternative hypothesis under the options button to "Not Equal," and we see the output below.

```
One-Sample Z
Test of mu = 40 vs. not = 40
The assumed standard deviation = 2.68


 N     Mean   SE Mean          95% CI          Z      P
30   41.3700   0.4893   (40.4110, 42.3290)   2.80   0.005
```

Comparing this with the one-tailed example (Section 2.4.7) we see that there are both similarities and differences.

- We now see "Test of mu = 40 vs. not = 40" instead of "Test of mu = 40 vs. > 40" (note that MINITAB had the second of these expressions, representing the one-tailed test, incorrectly — it should say "mu ≤ 40 vs. > 40").
- The mean, standard error of the mean, and z-statistic values are the same as before.
- The p-value is 0.005 instead of 0.003. In fact, there is a slight rounding error here — it is really exactly twice as big. This is because we have to add together the right tail $p(z > 2.80)$ and the left tail $p(z < -2.80)$.
- The confidence interval is given both an upper and a lower boundary, but the lower boundary is slightly smaller (i.e., further from the mean) than before (40.41 instead of 40.56 to 2 d.p.)

If we have all the data in the column "toothfishsamp," we can see the data histogram with the confidence interval (Figure 2.1), which shows both lower and upper bounds, and the fact that the lower bound is further away from the mean.

### 2.5.1 Why Is the Two-Tailed Confidence Interval Different from the One-Tailed?

If you look again at Section 2.4.6, you will see that the criterion we use to reject the Null hypothesis in the one-tailed case is the value of z such that the probability that it is exceeded is 5% or 0.05. Using Table C.1 (Appendix C), we found this value to be approximately 1.64. In terms of the confidence interval, we calculated the lower bound to be

$$mean - 1.64 \times s.e. = 41.37 - 1.64 \times 0.4893 = 40.57 \text{ (to 2 d.p.)}$$

In the two-tailed case, however, we need to find the value of z such that the right and left tails together add up to 5%. We can deduce this from the information in Table C.1, but it needs some thought. We already know that the right tail probability p(z > 1.64) is 5%. Because the z-distribution is symmetrical about its mean (0), it follows that if we could find the value of z such that the right tail probability is 2.5%, we would have the criterion value we need. So, in Table C.1, we need to find the value of z where p = 1 − 0.025 = 0.975. The closest we can get is somewhere between z = 1.95, where p = 0.9744, and z = 2.00, where p = 0.9772. In fact, we use the approximation 1.96. Thus, we now calculate the lower bound as

$$mean - 1.96 \times s.e. = 41.37 - 1.96 \times 0.4893 = 40.41 \quad \text{(to 2 d.p.)}$$

which agrees with the values shown in the MINITAB output.

We can see the determination of the one-tailed critical value for z in Figure 2.7.

### 2.5.2    What if the New Area Turned Out to Have Smaller Fish?

Suppose the sample of new catches turned out to have a mean weight of less than 40 t, say 38.63, for the sake of argument, with a standard deviation of 2.68. The conclusion we would draw would depend on the research hypothesis. If we were testing whether the mean length of the new fish differed from the historically established mean of 40 cm, we have a two-tailed test:

```
One-Sample Z: smallfish
Test of mu = 40 vs not = 40
The assumed standard deviation = 2.68
Variable    N    Mean   StDev SE Mean        95% CI              Z      P
smallfish  30 38.6267 2.6844  0.4893 (37.6677, 39.5857) −2.81 0.005
```

Apart from the fact that z is now negative,

$$z = \frac{\bar{x} - \mu}{\sigma / \sqrt{n}} = \frac{38.63 - 40}{2.68 / \sqrt{30}} = -2.81.$$

Apart from that, the calculations look much the same. If we look at the graphical representation of the test (Figure 2.8a), we see that the Null hypothesis mean (40) lies outside the confidence interval above the upper boundary. We therefore reject $H_0$ as before.

If, however, we test the original hypothesis, that the mean catch weight is *greater than* 40 t, then a different result emerges because it is now a one-tailed test:

```
One-Sample Z: smallfish
Test of mu = 40 vs > 40
The assumed standard deviation = 2.68

                                        95%
                                       Lower
Variable      N     Mean    StDev  SE Mean    Bound      Z      P
smallfish    30  38.6267   2.6844   0.4893  37.8218  −2.81  0.997
```

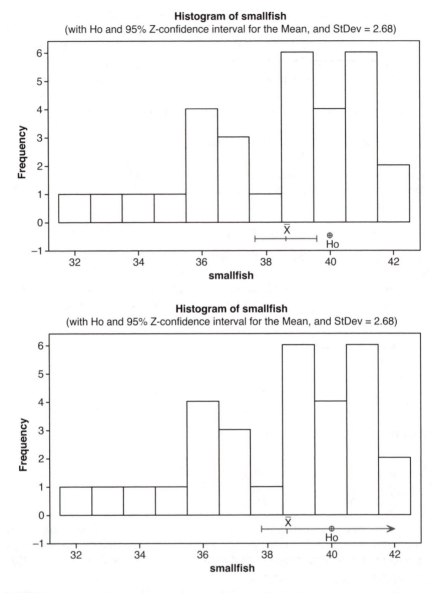

**FIGURE 2.8**
(a) Graphical representation of two-tailed z-test. (b) Graphical representation of one-tailed z-test.

At first sight it all looks the same; the summary statistics and even the z-value are the same, but now the p-value is nowhere near the significance criterion; in fact, it is 0.997, which means that if we reject the Null hypothesis on the basis of these data, we are *extremely likely* to make a Type I error. Looking at Figure 2.8b we realise that this is because the Null hypothesis mean 40 lies inside the confidence interval, which has no upper bound.

Although it is quite likely that the mean 38.62 is less than 40, this forms a part of the Null hypothesis, namely, that the mean fish length is less than or equal to 40 cm. Thus, in this situation, we should not reject $H_0$, but we would probably consider that we had done the one-tailed test "the wrong way around," meaning that we chose an inappropriate alternative or research hypothesis.

### 2.5.3 Tails and Tables

In Table C.1 we have given the probabilities for z values both above and below the mean (0). It is, in fact, quite common for z-tables to be given only for the part of the distribution above its mean, relying on the fact that it is symmetrical. Thus, if we want the left-tail value for z = 2.00, p = 0.0028 (as looked up in Table C.1), we could get the same value by looking up z = +2.0, p = 0.9772 in Table C.1, but then calculating 1 − 0.9772 = 0.0228 instead.

### 2.5.4 When You Can and Cannot Use a One-Tailed Alternative

On the face of it, a one-tailed test is tempting if you have already discovered the direction of the discrepancy between the means for the research and Null hypotheses because the one-tailed test is "more likely to give a significant result." Unless there are very specific reasons for having a directional research hypothesis (i.e., *less than*, or *greater than*, rather than *not equal to*) you must not use a one-tailed alternative. If you do, you are claiming that the probability of a Type I error, should the Null hypothesis in fact be true, is half as much as is really the case. In most computer programs and text-book treatments of the two-tailed test is the "usual" test (reflected in the fact that we had to explicitly choose the "greater than" alternative in Section 2.4.7). One-tailed alternatives are actually less common.

---

## 2.6 The Other Side of the Coin — Type II Errors

Given the apparent obsession with avoiding Type I errors, you might wonder why we settle for 5% as our rejection criterion. It means that one out of every 20 tests could be a false positive, which does not seem that stringent. Given the number of statistical tests we do, it might seem there was a significant risk that a Type I error would occur in practical data analysis. To see this point we need to consider another type of statistical error, the probability that we will fail to reject a Null hypothesis when in fact it is false. This is called a **Type II error** (Table 2.2):

**TABLE 2.2**

Type I and Type II Errors

|  | Test Not Significant | Test Significant |
|---|---|---|
| Null hypothesis true | Accept Null hypothesis when it is true | Reject Null hypothesis when it is true — **Type 1 error** |
| Null hypothesis false | Accept Null hypothesis when it is false — **Type II error** | Reject Null hypothesis when it is false |

You might think that there is a simple relationship between these two — e.g., that

$$p(\text{Type I error}) = (1- p \text{ (Type II error)}) \text{ WRONG!!}$$

but, unfortunately, life is not that simple. There is, however, a general negative relationship in the sense that the more stringent we make our Null hypothesis rejection criterion $\alpha$, the bigger the risk of not observing an effect which was really there. 5% or 1% turns out to be the best compromise, for most practical situations. Appendix B looks at this in more detail.

## 2.7   Recap — Hypothesis Testing

The procedure is completely general:

- Develop **research hypothesis** $H_A$. You must decide whether this is to be a directional (one-tailed) or nondirectional (two tailed) alternative.
- Deduce **Null hypothesis** $H_0$, which we will try to disprove. Be sure that you use the correct Null for the chosen research alternative.
- Collect data.
- Use data to calculate a **test statistic.**
- Use distribution of test statistic to find how likely the calculated value is *if the Null hypothesis is actually true.*
- Decide on criterion $\alpha$ for rejection of $H_0$ (what is an acceptable risk of a Type I error should $H_0$ happen to be true?)
- Decide whether to accept or reject $H_0$.

If possible, you should tattoo this on the inside of your eyelids. We will keep coming back to it.

Even so, you should beware of taking too mechanical an approach to the final decision. For example, a test that rejects $H_0$ at an $\alpha$ level of p = 0.049 is only slightly more convincing than one that rejects it at an α level of p = 0.051. How you would respond to this situation depends on the context — for example, if it was important to rule out an effect for some reason, a p-value as low as 0.051 might be worrying, but if you were looking for an important new effect, a p-value as high as 0.049 would not be very convincing. In the end it is a matter of informed judgment.

## 2.8 A Complication

Our use of the Normal distribution assumes that we have a large number of observations (usually taken to be more than 30) in our sample. We gave up on the worm example on the grounds that we had fewer than this, in fact only 10. The basic problem is that the use of the sample standard deviation as an estimate of the population standard deviation becomes progressively less reliable as the sample size decreases. It actually tends to underestimate the true population standard deviation, leading to a narrower confidence interval for the true mean. The problem with this is the increased probability of a Type I error. If we choose a criterion of z = 1.96 for the rejection of $H_0$, thinking that this gives us only a 5% chance of a Type I error if the Null hypothesis is true, the true risk will actually be greater than that if n < 30.

What is needed is an adjustment to the Normal distribution proportional to the amount of data we have, to bring the probability of a Type I error back to the 5% we want. This adjusted distribution is called **Student's t distribution.** Like the standard Normal distribution, it exists in table form (Table C.2, Appendix C), so, from the tables we can find out the appropriate value of *t* with which to multiply the standard error in order to obtain a 95% confidence interval corrected for the small sample size. The value of *t* gets larger as the number of observations gets fewer, correcting the tendency of z to be too narrow in this situation. As with the Normal distribution, the value of *t* also gets larger as the amount of confidence we require gets larger. So, for example, with 30 worms in the sample and only 95% confidence required, *t* = 2.04, but if 99% confidence is needed and the sample is only 10 worms, *t* increases to 3.35 (Table 2.3).

One catch to beware of is that t-tables are normally printed with two-tailed probabilities as the column headers. If you want to do a one-tailed t-test, then you have to divide the column heading (p-value) by 2. If you forget which way this works, remember that the two-tailed 95% critical value in the z-distribution is 1.96, which is the same number as you find in the 5% column of Table C.2 for infinity degrees of freedom. The one-tailed 5% critical value is 1.64 (Figure 2.5), and this is the value we find at the bottom of the column for p = 10% in Table C.2. For large numbers of degrees of freedom, the z and t distributions converge.

**TABLE 2.3**

Selected Values of Student's t Distribution

| Number of Worms | Degrees of Freedom | 95% Confidence | 99% Confidence |
| --- | --- | --- | --- |
| 10 | 9 | 2.26 | 3.35 |
| 20 | 19 | 2.09 | 2.84 |
| 30 | 29 | 2.04[a] | 2.76 |

[a] This is quite close to the value of 1.96, which comes from the Normal distribution; in fact it is just under 5% out (0.08/2.0 = 0.04), so when n > 30 the Normal distribution can be used instead of Students t. In many practical situations an approximate 95% confidence interval is given by the mean plus or minus twice its standard error, provided our sample contains more than 30 individuals.

## 2.9   Testing Fish with t

Testing our toothfish sample using the t-statistic instead of the z-statistic could not be easier. The calculation of the test statistic is, in fact, the same, except that now we do not have to make the claim that the sample standard deviation s is a good estimator of σ; we just use s anyway.

So we calculate

$$t = \frac{(\bar{x} - \mu)}{s/\sqrt{n}} = \frac{(41.367 - 40)}{2.68/\sqrt{30}} = 2.79.$$

Because the calculations involved are exactly the same as for the z-test, the value of the test statistic is 2.79 again, so all we have to do is look up the different tables (Table C.2). However, when we do, we find a new column that was not present in the z-tables — entitled "degrees of freedom." In this case there are $n - 1 = 29$ degrees of freedom.

There is a slight complication caused by the fact that we are conducting the test as a one-tailed test. As explained in Section 2.8, the critical value we want is in the column headed 10 (representing the rejection probability 10%, or p = 0.10) for both the tails, so for the left tail, only the p-value is 10/2 = 5%, which is what we want. We choose the appropriate row for the degrees of freedom. There is a slight problem here in that there is no row for 29 d.f. What we should do here is choose the next lowest row, i.e., 24, for degrees of freedom, i.e., 1.71. The reason for this slightly arcane procedure is that we would rather pick a *t* that was a bit too big than one that was too small, making the test conservative with respect to the chance of a Type I error. (We could use MINITAB to look it up directly, see Appendix A, Section A.8.3 — the true value is 1.699). The calculated value for *t* is comfortably greater than 1.71, so these manipulations were not really necessary, and we can easily reject H₀ at the 5% level using the same logic as we used for the z-test in Section 2.4.6.

## 2.10 MINITAB Does a One-Sample t-Test

We ask for a one-sample t-test with the command **Stat>Basic Statistics>1-Sample t Samples in columns**: toothfishsamp **TestMean**: 40. Note that here we did not have to supply the value of the population standard deviation because the t-test uses the sample standard deviation that can be calculated from the data, but we must set our alternative hypothesis to "greater than" to make the test one-tailed. Its also a good idea to ask for the histogram of the data, which shows the hypothesis test graphically in the same way as we saw for the z-test:

```
One-Sample t: toothfish
Test of mu = 40 vs. > 40

                                        95%
                                      Lower
Variable    N    Mean   StDev   SE Mean   Bound     T      P
Toothfish   30   41.3667 2.6844  0.4901  40.5339  2.79  0.005
```

As with the z-test, we use a one-tailed alternative, that catches are more than 40 t in weight, so we see only the lower bound of the confidence interval. MINITAB gives the exact probability for the calculated value of t (2.79) with 29 d.f.; it is 0.005, well below the normal rejection criterion of 0.05.

It is worth making a point-by-point comparison between the output for the t-test and that of the z-test on the same data. The summary statistics N, mean, and standard error are all the same, but the t-test has calculated the standard deviation from the data. The 95% confidence interval calculated from the t-distribution is very slightly *wider* and, consequently, the p-value is *larger* (0.005 instead of 0.003) This is what we expect — the z-test has given a confidence interval that is slightly too narrow, and consequently, it underestimated the probability of the unlikely event that would have to have occurred if $H_0$ was really true. Although the differences here are trivial, the discrepancy will get progressively larger as the sample size decreases. Compare the 5% column value for infinity degrees of freedom (1.96, the same as the z-distribution) with that for 2 d.f. (4.30). The graphical representation of the test (Figure 2.9) looks similar to the one generated by the z-test, but now the confidence interval calculated is based on the critical value in the t-distribution and the sample standard deviation.

## 2.11 95% CI for Worms

Returning to our sample of 10 worms from Chapter 1, what can we deduce about the likely true mean of the population from which the sample was

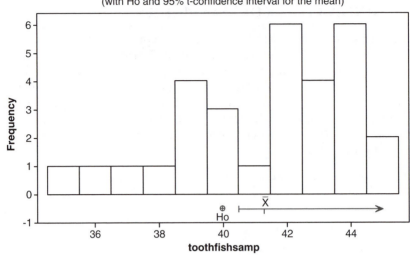

**FIGURE 2.9**
Graphical representation of one-tailed t-test.

drawn? We ask MINITAB to do the sums for us using **Stat>Basic Statistics>1-sample t**; if we do not give a test mean, it just gives the summary statistics and the confidence interval for the mean. As we do not have any views about how long worms ought to be, we *must* remember to set the alternative hypothesis here to "equal."

```
One-Sample T: worms
Variable    N     Mean   StDev   SE Mean        95% CI
Worms      10  10.0000  2.1731   0.6872  (8.4455, 11.5545)
```

The corresponding graphical output is shown in Figure 2.10. Because this is a two-tailed confidence interval we have both an upper and lower bound.

In fact, this seems astonishingly precise. We can be 95% certain that, given these data, the true mean lies in the range 8.45 to 11.55 cm. If we drew another sample of 10 worms, of course, the mean, standard deviation, and hence standard errors would all be different, but for 95% of samples the confidence interval calculated from the sample mean and standard deviation would contain the true mean.

## 2.12 Anatomy of Test Statistics

We have met with two test statistics in this Chapter. They were called "z" and "t," after the names of the distributions they follow when the Null

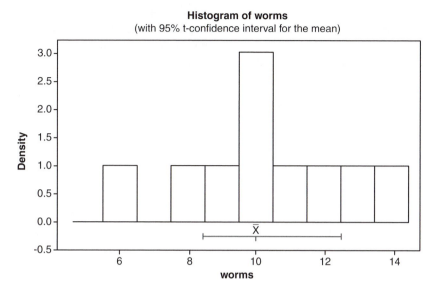

**FIGURE 2.10**
Graphical representation of two-tailed t-test.

hypothesis they are being used to test is in fact true. Looking again at the schematic formulae

$$z = \frac{(\bar{x} - \mu)}{\sigma/\sqrt{n}} \text{ and } t = \frac{(\bar{x} - \mu)}{s/\sqrt{n}}$$

we see a pattern beginning to emerge. So far, test statistics seem to share the form

$$test\_statistic = \frac{discrepancy}{SE(discrepancy)}$$

where "discrepancy" is the difference between the value of the statistic (in this case a mean) we expect if the Null hypothesis is true, and what we actually observe. One helpful thing about this insight is that it gives us a mnemonic for when we forget how to read our statistical tables: the bigger the absolute value (i.e., ignoring the sign) of the test statistic, the less likely it is to have come about from a situation where the Null hypothesis is true.

You have now encountered some very important concepts (as well as learning how to measure worms). Do not be surprised if you do not understand or remember them after just one reading. Most people find that they need to come across these ideas many times before they are comfortable with them. The best way to understand the subject is to have some observations of your own that you wish to summarise. Then you will be motivated to put these methods to good use. If you are puzzled, refer back to the appropriate section and reread it. Reading about statistical methods cannot be done at the same fast pace as reading a light novel or a science fiction book — but it will repay your attention if done with care.

# 3

## Comparing Things: Two Sample Tests

... when I call for statistics about the rate of infant mortality, what I want is proof that fewer babies died when I was Prime Minister than when anyone else was Prime Minister.

**—Winston Churchill**

In Chapter 2 we looked at some theoretical properties of samples and how we can deduce properties about the populations from which they were drawn. We could also ask questions about whether a sample was likely to come from a population with some known property, and on this basis we saw that we can test hypotheses about populations. In this chapter we extend this idea to the more common situation where we want to compare two sets of measurements to see if they are the same. We will find that basically the same ideas are used.

### 3.1  A Simple Case

The function of sleep is one of the enduring puzzles of Biology. One idea is that it serves some recuperative function, and therefore we sleep longer after a day of physical exercise. Another view is that it is to do with processing information acquired the previous day, so we might expect to sleep longer after a day of mental activity. In an experiment to test these ideas against one another, 36 students were asked to time their continuous sleep after a day of strenuous physical activity, and again after a day spent attending classes. Exactly how a study like this should be designed is discussed further in Chapter 4, but we would obviously pick the students at random, possibly have only one sex, or else equal numbers of males and females. It would also be a good idea to do the exercise day first for half the subjects and second for the others. The data collected are shown in Table 3.1. How do we test the hypothesis that sleep is affected by exercise?

**TABLE 3.1**

Number of Hours of Sleep Experienced by 36 Students
Following Exercise and Attendance at Classes and
Differences in Sleep Following These Activities

| Exercise | Classes | Difference |
|----------|---------|------------|
| 8.38  | 7.46  | 0.92 |
| 9.10  | 8.31  | 0.79 |
| 9.45  | 8.78  | 0.67 |
| 5.98  | 5.70  | 0.28 |
| 10.52 | 9.63  | 0.89 |
| 9.35  | 8.65  | 0.70 |
| 10.86 | 11.24 | 0.38 |
| 7.97  | 6.70  | 1.27 |
| 6.69  | 6.18  | 0.51 |
| 7.40  | 6.40  | 1.00 |
| 3.57  | 3.97  | 0.40 |
| 11.11 | 13.14 | 2.03 |
| 4.12  | 4.05  | 0.07 |
| 9.66  | 9.07  | 0.59 |
| 4.86  | 5.31  | 0.45 |
| 10.40 | 9.25  | 1.15 |
| 10.68 | 10.60 | 0.08 |
| 8.55  | 7.82  | 0.73 |
| 9.16  | 8.45  | 0.71 |
| 9.37  | 8.69  | 0.68 |
| 8.13  | 6.92  | 1.21 |
| 10.45 | 9.38  | 1.07 |
| 9.28  | 8.55  | 0.73 |
| 5.75  | 5.44  | 0.31 |
| 8.57  | 7.86  | 0.71 |
| 8.66  | 7.92  | 0.74 |
| 8.82  | 8.29  | 0.53 |
| 6.49  | 5.82  | 0.67 |
| 10.25 | 9.07  | 1.18 |
| 8.18  | 7.40  | 0.78 |
| 6.56  | 5.88  | 0.68 |
| 6.18  | 5.72  | 0.46 |
| 4.73  | 5.26  | 0.53 |
| 10.49 | 9.46  | 1.03 |
| 5.79  | 5.56  | 0.23 |
| 8.09  | 6.92  | 1.17 |

## 3.2  Matched-Pairs t-Test

Each subject was tested both after exercise and after a day of mental activity
(attending lectures), so the obvious thing to do is compare the duration of
sleep under each condition for each subject by simply subtracting one duration from the other. This is shown in the third column of Table 3.1. If there

were no tendency for either exercise or mental activity to be followed by longer sleep durations, we would expect that the difference column would on average be 0. So, the hypothesis test is simply testing whether the difference is likely to be zero. Let us see how this works in practice.

1. Define the research hypothesis. Duration of sleep is different after physical exercise compared with mental activity. Because we can test each subject under both conditions, this can be represented by saying that we expect that the difference between sleep duration after exercise and sleep duration after mental activity is not equal to zero (on average). Note that we have no expectation about whether physical exercise or mental activity (attending lectures) will generate more sleep.

2. Deduce the Null hypothesis. If exercise has no effect, then the mean of the difference in sleep duration between the two conditions should be zero.

3. Collect the data — as discussed previously (Table 3.1).

4. Calculate the probability of the data if the Null hypothesis is true. This is exactly similar to the procedure for testing whether the mean of a single sample is equal to zero or not, devised in Section 2.8. The summary statistics for difference are given in Table 3.2.

We use these to calculate the standard error of the mean difference:

$$s\Big/\sqrt{n} = sample\_sd\Big/\sqrt{samplesize} = 0.647\Big/\sqrt{36} = 0.1078$$

and the test statistic is now

$$t_{35d.f} = \frac{difference}{se(difference)} = \frac{0.521}{0.1078} = 4.83 \,.$$

To find the probability we look in Appendix C, Table C.2. If the degrees of freedom we want are not represented, we use the *next lowest* d.f. as this is conservative and minimises the risk of a Type I error. So, using 30 d.f., we find that 4.83 comfortably exceeds the value in the $p = 5\%$ column (2.04) and indeed is greater than the critical value for the lowest probability shown ($p = 0.1\%$, $t_{crit} = 3.29$).

**TABLE 3.2**

Summary Statistics for Difference in Table 3.1

| Sample Size | Mean | Standard Deviation | Standard Error of Mean |
|:---:|:---:|:---:|:---:|
| 36 | 0.521 | 0.647 | 0.1078 |

5.  Decide on a criterion for rejection of the Null hypothesis. If we adopt the conventional 5% criterion, then clearly the probability of getting t = 4.83 with 35 d.f. is much less than that.

6.  We therefore reject the Null hypothesis that exercise does not make a difference to sleep duration.

Although the hypothesis test was not specific about the direction in which the difference would be (our alternative hypothesis was just the two treatments were not equal), we can now ask in which direction the effect actually goes. Here we make use of the 95% confidence interval for the mean difference. Using the method devised in Section 2.7, the interval is defined as

$$95\% \ CI = difference \pm t_{35d.f.} \times s.e. \ (difference).$$

We agreed to use the critical value from the t tables for 30 d.f. (2.04) as we do not have a value for 35 d.f. We realise that our calculation is slightly wrong, but by using the larger value of t, we make the confidence interval a bit too wide rather than a bit too narrow. Being conservative ensures that the probability of making a Type I error if the Null hypothesis is true is no more than 5%; in fact, it will be slightly less than that. We, therefore, calculate the interval as

$$95\% \ CI = 0.521 \pm 2.04 \times 0.1078 = 0.521 \pm 0.220$$

$$= [0.301, 0.741]$$

Because the difference was calculated as *exercise–classes* and the confidence interval lies above zero rather than below it, we can conclude that the subjects slept longer after physical exercise than they did after a day attending classes.

## 3.3  Another Example — Testing Twin Sheep

A farmer is wondering whether to change the food supplement he gives to his lambs after weaning. Sheep have the very useful habit of producing twins, which suggests a simple but effective experiment. Take a number of pairs of twin sheep and feed one of each pair on the new diet, and the other on the old diet. The data from such an experiment are shown in Table 3.3. As before, what we are doing here is testing whether the mean difference between the growth rates of the two twins is greater than zero.

There are two differences to notice with this analysis compared to Section 3.2:

1.  We cannot raise a lamb both on the old and on the new diet, so we cannot use each subject twice as we did in the sleep experiment. Instead, we must find a plausible way of selecting pairs of subjects

**TABLE 3.3**

Weight Gain (kg per Unit Time) of Seven Pairs of Lambs on Old and New Diet

| Old | New | Difference |
|-----|-----|------------|
| 3.4 | 4.5 | 1.1 |
| 3.9 | 4.8 | 0.9 |
| 4.2 | 5.7 | 1.5 |
| 4.5 | 5.9 | 1.4 |
| 3.6 | 4.3 | 0.7 |
| 2.9 | 3.6 | 0.7 |
| 3.2 | 4.2 | 1.0 |

that are likely to be comparable in most respects other than the treatments to which they are subjected. In this case, the convenient fact that sheep often produce twins solves the problem, but life is often less simple. In medical trials, for example, it is common to try to **match** pairs of subjects for factors that might have a bearing on the outcome of their disease, such as age, gender, and socioeconomic status, and then randomly assign one of the pair to each treatment. Hence the name of the test — "Matched Pairs."

2.  In this example we will investigate the research hypothesis that the new diet is *better than* the old one, rather than just asking if it is *different*, as we did in the sleep example. If the new diet generally produces a higher growth rate, then the difference for each pair *newdiet–olddiet* will on average be positive, so we conduct this as a **one-tailed test** with the alternative hypothesis that the mean difference is greater than 0.

To illustrate this analysis we show the output from the MINITAB paired t-test:

```
Paired t-Test and CI: New, Old
Paired t for new—old

            N     Mean     StDev   SE Mean
    New     7   4.71429   0.82750   0.31277
    Old     7   3.67143   0.56484   0.21349
Difference  7   1.04286   0.31547   0.11924

95% Lower bound for mean difference: 0.81116
t-Test of mean difference = 0 (vs > 0): t-value = 8.75
p-value = 0.000
```

The test statistic here is t with 6 degrees of freedom, calculated as

$$\frac{mean\_of\_differences}{s.e.\_of\_differences} = \frac{1.04286}{0.11924} = 8.75 \text{ (to 3 s.f.)}.$$

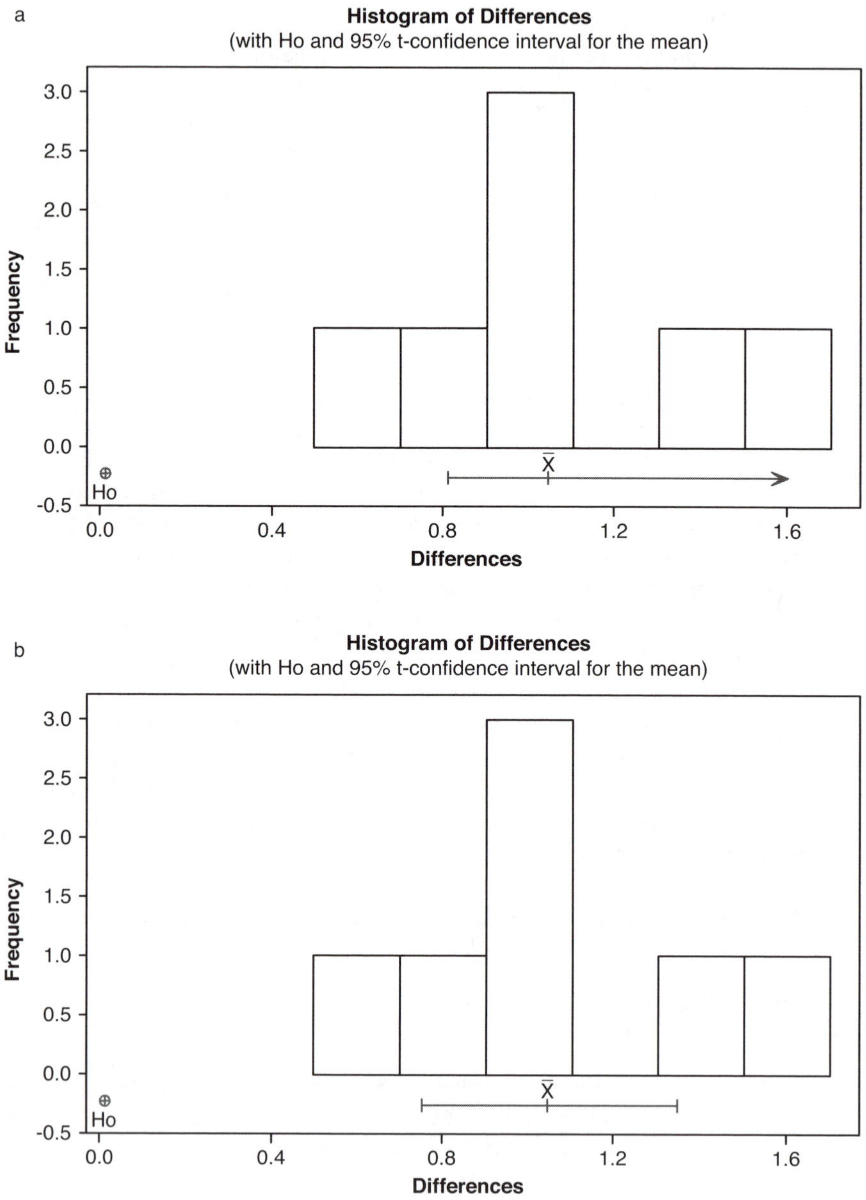

a

b

**FIGURE 3.1**

(a) Graphical representation of paired t-test of data in Table 3.3. (b) The paired analysis for sheep weight carried out as a non-directional (two-tailed) test, for comparison. Note that we now see an upper bound for the confidence interval of the mean, and that the lower bound is closer to the Null hypothesis value (0.751 instead of 0.811). See text for fuller explanation.

This is considerably larger than the 5% critical value for 6 d.f. in Table C.2, which is 1.94 (remembering that we must look in the 10% column because this is a one-tailed test and we are doubling the area we are looking at because we are only interested in one tail. Look again at Section 2.4 if you have forgotten about this). The test is also illustrated in Figure 3.1a, which shows the histogram of the differences together with the confidence interval for the mean difference; the Null hypothesis value (0.0) lies well outside this interval to the left. Note that the upper bound of the confidence interval does not exist, so it has not been calculated. The lower bound is

$$1.04286 - 1.94 \times 0.1192 = 0.811 \text{ to 3 s.f.}$$

We can therefore be quite confident that the new diet gives higher weights than the old one.

## 3.4 Independent Samples: Comparing Two Populations

Unfortunately, we cannot rely on being able to match up subjects as required for the matched pairs t-test. It is quite common to have two samples, one from each of two groups that differ in some respect in which we are interested. To take an obvious example, if we were interested in gender differences, we could not test the same individuals both as females and as males because each subject must be one or the other. The research hypothesis in this case would be that the means of the two samples differ; another way to express this is to say that they are random samples from two different populations. From this we may deduce the Null hypothesis, which is that they come from the same population and so have the same mean. If we are satisfied that the evidence from our sample data is strongly against this idea, we reject it and instead accept $H_A$ that the samples come from populations with different means.

### 3.4.1 The Two-Sample t-Test

This test is also known as an **independent samples t-test**. Like the paired t-test, we test the Null hypothesis by looking at a difference, but now it has to be the difference between the two group means. The further the difference between the means is from zero, and the less the overlap in the individual values around each of them, the more likely it is that the samples come from populations with different means. In that case, we can reject the Null hypothesis at a given level of confidence (e.g., $p = 0.05$), just as we do for other hypothesis tests.

To get a feel for the difference, we should plot data. There are several ways of doing this, but what we need is a plot that shows the location of the means and the amount of scatter around them. An "individual values plot" is satisfactory for small amounts of data.

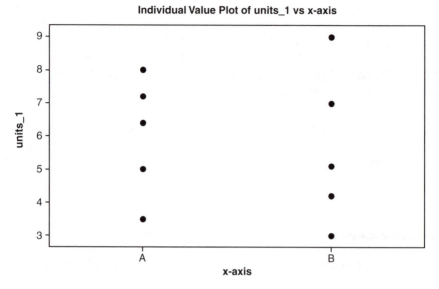

**FIGURE 3.2**
Two sets of observations overlap completely.

- If the two sets of observations overlap each other a great deal, it may be so obvious that the Null hypothesis cannot be rejected that it is pointless carrying out a t-test (Figure 3.2).
- If the observations in one sample are all clearly separated from those in the other sample, it may be so obvious that the Null hypothesis is false that a t-test is superfluous (this assumes that you have a reasonably large number, say 5 or more, of observations in each group). (Figure 3.3)

Often our data are intermediate, with some overlap between the two samples (Figure 3.4). Here we need to carry out a statistical test.

## 3.5  Calculation of Independent Samples t-Test

Suppose in our sheep diet example that, instead of seven pairs of twin sheep, we have 14 different, unrelated, individuals, seven of which were selected at random to receive a *new* diet, whereas the other seven were given the *old* diet. It no longer makes sense to find the differences in weight gain for the pairs of numbers in the data table because the animals in the two groups are not related to each other in any way. Instead, we calculate the mean value for each group. We wish to judge the magnitude of the difference between the two means compared to a difference of zero, which is what we expect if the diets have the same effect.

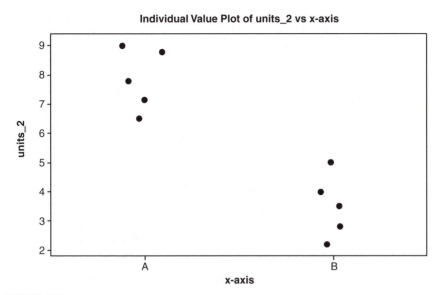

**FIGURE 3.3**
Two sets of observations do not overlap at all. (Jitter added along the x-axis to avoid the problem of points overlaying each other within a sample.)

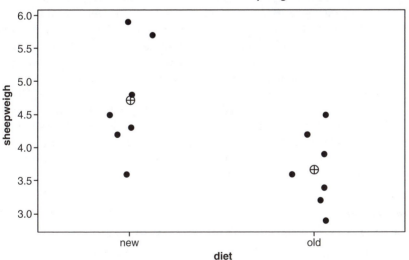

**FIGURE 3.4**
Two sets of observations partially overlap. Note (a) the jitter as in Figure 3.3, (b) the means are shown for clarity.

1. Define the research hypothesis. Sheep grown on the new diet will have *greater* weights than those grown on the old diet, making this a one-tailed test. We will work out the difference between the means as *mean(new) – mean(old)*.

2. Deduce the Null hypothesis. If the new diet is no better than the old, then the difference between the means for the two groups should be zero or less than zero.

3. Collect the data.

4. Calculate the probability of the data if the Null hypothesis is true. The test statistic is t calculated as

$$t = \frac{(mean(new) - mean(old)) - zero}{se(difference)}$$

The probability of this value of t if the Null hypothesis is true is determined from Table C.2 with $n - 2 = 12$ degrees of freedom. Because the test is one-tailed, we use the column for the 10% critical values (see Chapter 2, Section 2.5, again if you have forgotten about this).

5. Decide on a criterion for rejection of the Null hypothesis.

6. Decide whether to accept or reject the Null hypothesis that diet does not make a difference to sheep weight.

Compare this with the procedure for the matched pairs test, and you can see that the idea is the same.

The principle of the test — calculating a test statistic from the discrepancy between the expected difference in means (0) and what we actually observe, (4.714 − 3.671 = +1.043) — is just the same as before. The crucial difference is that whereas the standard error of the differences between paired observations is easy to calculate, the standard error for the difference in the means of two independent samples is more of a problem. For the moment we give the formula for this without justification, but we will return to it in Chapter 6. The trick is to calculate something called the "pooled variance" estimate $s^2{}_p$, which appears in Equation 3.3 (in Box 3.2). Equation 3.1 shows you how to do it, and any sensible person would get a computer to do the calculations, but there is a calculator method relying on the sample standard deviation button that you will find on a statistical calculator.

## Box 3.1 A Simple Method of Extracting Sums of Squares

It is our general contention in this book that calculation is a job for computers, not people, but working through the calculations manually can be helpful at the start. The direct way to get a sum-of-squared deviations is, of course, to work out the deviations, square each one, and add them all up. So, for the old diet (using the figures in Table 3.3), we would first have to calculate the mean, which turns out to be 3.67 (to 3 s.f.), and then:

$$(3.40 - 3.67)^2 + (3.9 - 3.67)^2 + \ldots$$

for all seven sheep on the old diet. Quite apart from being extremely tedious, it is quite likely we would make a mistake!

A better method relies on the fact that most calculators have a "standard deviation" program. In Chapter 1 we saw that, in mathematical terms, the standard deviation is defined as

$$s = \sqrt{\frac{\sum_{i=1}^{n} (x_i - \bar{x})^2}{(n-1)}}$$

where the term

$$\sum_{i=1}^{n} (x_i - \bar{x})^2$$

is precisely the sum-of-squared deviations about the mean we are after — let us write it as *SSX* to make things simpler. We then use a simple bit of algebra to show that

$$s = \sqrt{\frac{SSX}{(n-1)}}$$

$$s^2 = \frac{SSX}{(n-1)}$$

$$s^2 \times (n-1) = SSX$$

To use this on the calculator, all you need to do is:

- Set the calculator to "statistical" mode.
- Enter all seven observations for the group (Table 3.3).
- Press the sample standard deviation button $\sigma(n-1)$.
- Square it to get the sample variance.
- Multiply this by n − 1 to get the total sum-of-squares.

For the old diet the answer should be 1.9143.

## BOX 3.2 Equations for the Two-Sample t-Test

The pooled variance is worked out from the sum-of-squared deviations about the mean for the data of each group, which can be obtained from the standard deviations for each group using the method proposed in Box 3.1. This gives us

$$s_p^2 = Pooled\ variance = \frac{SSX_1 + SSX_2}{(n_1 - 1) + (n_2 - 1)} = \frac{(n_1 - 1)s_1^2 + (n_2 - 1)s_2^2}{(n_1 - 1) + (n_2 - 1)} \qquad (3.1)$$

The standard error of the difference may be calculated as

$$SED = \sqrt{\left(\frac{s_p^2}{n_1} + \frac{s_p^2}{n_2}\right)} = \sqrt{\left(s_p^2\left(\frac{1}{n_1} + \frac{1}{n_2}\right)\right)} \qquad (3.2)$$

and, finally, we obtain the t-test statistic for the Null hypothesis that the difference between the means is zero by

$$t = \frac{difference - 0}{SED} = \frac{(\bar{x}_1 - \bar{x}_2) - 0}{\sqrt{s_p^2\left(\frac{1}{n_1} + \frac{1}{n_2}\right)}} \qquad (3.3)$$

To calculate $s_p^2$ we have to find the "sum of squared deviations of the data about the mean" for each of the samples separately. The manual procedure is therefore as follows:

- Calculate the sum-of-squares of the observations in the old diet group and add this to the sum-of-squares of the observations in the new diet group. Making use of the trick explained in Box 3.1, obtain the standard deviation from the calculator ($\sigma_{n-1}$), square it to obtain the variance, and multiply by $(n - 1)$ to find the sum-of-squares.

  $(0.8275)^2 \times 6 = 4.1086 =$ new diet group sum-of-squares

  $(0.5648)^2 \times 6 = 1.9143 =$ old diet group sum-of-squares

  Adding together: $4.1086 + 1.9143 = 6.0229$
  $=$ total of both sums-of-squares

- Divide the result by the sum of the degrees of freedom for new diet group and old diet group:

  $(n_1 - 1) + (n_2 - 1) = 6 + 6 = 12$

  $6.0229/12 = 0.5019 =$ the pooled variance

- Then the standard error of the difference (Equation 3.2, Box 3.2) is obtained by multiplying the pooled variance by 1/7 plus 1/7 (because 7 is the number of sheep in each group) and taking the square root:

$$SED = \sqrt{0.5019 \times \left(\frac{1}{7} + \frac{1}{7}\right)} = 0.37868$$

- Finally, the t ratio (Equation 3.3, Box 3.2) is:

$$t = \frac{(difference - H_0)}{SED} = \frac{(1.043 - 0)}{0.37868} = 2.75$$

The positive sign indicates that the new diet has the greater mean (as proposed by the research hypothesis). We consult the t-tables to see whether 2.75 is greater than the critical value for the combined degrees of freedom, i.e., 12 d.f. Because it is a one-tailed test, we must look in the P = 10 column of Table C.2, row d.f. = 12 to find the critical value for $\alpha = 0.05$ (see Section 2.8 if you have forgotten about this).

As our calculated t is bigger than the table value of 1.78, we can be reasonably confident that the difference between the population means **is greater than** zero. Formally, this means that we can reject the Null hypothesis (of no positive difference) with 95% confidence. If the groups really do not differ in the predicted direction, we would get a discrepancy as big as this on less than 5% of occasions.

## 3.6 One- and Two-Tailed Tests — A Reminder

It is probably more usual to ask the question "Is there any evidence that the two treatments are *not the same?*" In other words, we would be equally interested in whether the new diet resulted in either bigger **or** smaller sheep than the new diet. This is known as a two-tailed test because the mean difference could lie in either "tail" of the distribution, i.e., away from the zero in the middle. The t-tables are usually presented as we have printed them in Appendix C, Table C.2, with the p-value listed as the appropriate value for **a two-tailed test**.

If we did decide that the sheep diet analysis should be a two-tailed rather than a one-tailed test, we would find the appropriate t-value by looking down the column headed 5. This means that there is 2.5% in each tail, giving 5% in total (i.e., p = 0.05); then answer with 12 d.f. is 2.18. Beware, other sets of tables may not behave this way. If you are unsure, you can check if the value for infinite degrees of freedom is 1.96. If it is then the numbers are 5% critical values for a two-tailed test because with infinite degrees of freedom, the t-distribution is identical to the standard Normal distribution (Appendix C, Table C.1).

## 3.7   MINITAB Carries Out a Two-Sample t-Test

First, we look at the data again. Notice that we have rearranged it into a single column for the weights and another specifying which diet was used (see Appendix A, Section A.7). Each row therefore represents a "case," underlining the fact that the weights in the "old" diet group have no relationship to the weights in the same row of the data set for the "new" diet that we used in the matched pairs analysis.

```
Data Display
Row      Sheepweigh      diet
  1             3.4       old
  2             3.9       old
  3             4.2       old
  4             4.5       old
  5             3.6       old
  6             2.9       old
  7             3.2       old
  8             4.5       new
  9             4.8       new
 10             5.7       new
 11             5.9       new
 12             4.3       new
 13             3.6       new
 14             4.2       new
```

Then we ask for a two-sample t-test (**Stat>Basic Statistics>Two Sample t Samples in one column Samples**: sheepweigh **Subscripts**: diet). Note that we MUST remember to check the **assume equal variances** box here. Because we will conduct the test as a one-tailed test, we must select the alternative hypothesis as "Greater than" under the Options button N.B.: There is a quirk here; although we put the old diet rows in first, MINITAB orders the treatments alphabetically. Thus, whether we like it or not, it will calculate the difference between the means as

$$mean(new) - mean(old) \text{ (because "n" becomes before "o").}$$

```
Two-Sample t-Test and CI: sheepweigh, diet
Two-sample t for sheepweigh

diet    N    Mean    StDev    SE Mean
 new    7   4.714    0.828      0.31
 old    7   3.671    0.565      0.21

Difference = mu (new)  > mu (old)
Estimate for difference:   1.04286
```

```
95% Lower bound for difference:   0.36793
t-Test of difference = 0 (vs >): t-Value = 2.75  p-Value = 0.009
d.f. = 12
Both use Pooled StDev = 0.7085
```

Note that MINITAB refers to the population means as "mu" (mu(new) and mu(old)). This is the sound of the Greek letter $\mu$ that is used to represent the population mean(s) that we have estimated by the *sample* means.

It concludes that we can reject the hypothesis that the new diet produces sheep that are no heavier, or even lighter, than those on the old diet. The p-value is 0.009 and, as always, this is the probability that data as extreme as this would occur if the Null hypothesis was true. As we saw in the one-tailed matched pairs test, only the lower bound of the confidence interval is printed. It shows that the lowest likely value (with 95% confidence) for the difference in means is 0.36973.

---

## 3.8   Pooling the Variances?

We insisted both in the manual and the computer calculations that you should pool the variances when carrying out a two-sample test. Why do we do this? When is it legitimate to do so? And what happens if it is not legitimate?

### 3.8.1   Why Pool?

Recall that the Null hypothesis is that the two samples are drawn from the *same* population. Although we may think this unlikely, after plotting the data for example, we proceed on the basis that it is true because this is the mathematical basis on which we can calculate the probability of getting means as different as we actually find them to be if $H_0$ is true.

If it was true that the samples come from the same population, then each of the two sample means is a separate estimate of the *same* population mean, and each of the sample standard deviations is a separate estimate of the *same* population standard deviation $\sigma$. If we put the two samples together, we could get a better estimate of $\sigma$ by combining the information from both of the samples. If you look carefully at the equations for calculating the pooled variance, you can see that it is the mean of the combined sample variances, weighted by the sample sizes.

### 3.8.2   When to Pool?

It is only sensible to form a pooled estimate of the variance when the sample variances (s.d.$^2$) are not too far apart. Even if the sample means are not

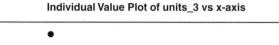

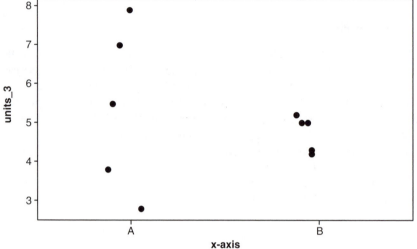

**FIGURE 3.5**
Two sets of observations have unequal variances.

different, the variation in weights about the mean might be. If the two variances are very different, this is evidence in itself that the populations behave differently, i.e., the new diet is having a different effect and there is little point in continuing with the comparison (e.g., Figure 3.5). We can test for this by dividing the bigger variance by the smaller one. As a rule of thumb, if this ratio is more than about 2.5, then we have evidence that the variances are significantly different. To test it formally, we need yet another test statistic, which we will meet in Chapter 6.

### 3.8.3   What if the Variances Are Different?

Opinion is divided about this, so if you run into this problem, you should consult a statistician. Some people advocate using an "unpooled" version of the t-test. This is implemented in most computer packages, including MINITAB. The principle of the test is the same, but the calculation of the t-statistic becomes:

$$t = \frac{(\bar{x}_1 - \bar{x}_2) - 0}{\sqrt{\left(\dfrac{s_1^2}{n_1} + \dfrac{s_2^2}{n_2}\right)}}$$

A major difficulty arises over the degrees of freedom for this test. They are definitely not $n_1 + n_2 - 2$. A conservative approach if doing the test by hand is to use the smaller of $n_1 - 1$ and $n_2 - 1$. MINITAB uses a complicated formula that often produces fractional d.f. (such as d.f. = 4.52), which requires explanation beyond the scope of this book (see Howell 2001).

Some authorities doubt the validity of the unpooled t-test altogether on the grounds that it violates one of the general assumptions behind the calculation of the distribution of the test statistic on the basis that the Null hypothesis is true. These people advocate first attempting a mathematical transformation of the data to try to produce similar variances for the two groups. This is discussed by Grafen and Hails (2002) at some length. If the transformation route fails, the use of a "nonparametric" test, in this case the Mann–Whitney test, is recommended. This is discussed in detail in Chapter 10.

# 4

---

## *Planning an Experiment*

---

> Organizing is what you do before you do something, so that when you
> do it, it is not all mixed up.

**A.A. Milne**

This chapter is split into two sections and deals with how to design an effective experiment. First, we will discuss the principles of sampling to show how different sampling regimes affect the results we obtain. Then, we will put this knowledge to use by discussing how to design your own experiment.

---

## 4.1 Principles of Sampling

### 4.1.1 First, Catch Your Worm!

In Chapter 2, we measured ten worms and considered how to describe and analyse them. How we actually obtain these data is not a trivial matter. There are practical as well as statistical issues. Let us consider the challenge.

First, there are practical problems: if we go out during the day, worms will be buried, and we will need to dig them up. How deep should we go? Also, we run the risk of cutting some of them, which we would then not be able to use. Would there be a bias towards cutting and discarding longer worms? It would probably be better to go out on a wet night with a torch and collect them into a bucket while they are on the surface. Is there a bias in which worms come up? All of these practical considerations might influence which individuals we pick.

How do we decide which ten worms to pick up from the many thousands available? If we collect them from the footpath because it is convenient, we might have individuals that can move through compacted soil and might therefore be smaller than average. We want a sample of ten worms that is *representative* of the *population* of thousands of worms in the field. Deciding on a method to pick ten worms is called **sampling**. This chapter will describe alternative methods of sampling, together with their strengths and weaknesses, and then discuss the parts of a good experimental design.

### 4.1.2   The Concept of Random Sampling

We have already discussed this in some detail in Chapter 1, Section 1.2.4. The purpose of random sampling is to ensure that each individual in the population of interest has an equal chance of being selected (no bias towards certain individuals). We usually take a random sample (in this case, of ten observations). This may be unrepresentative because, by chance, no observations have been selected from a distinctive part of the population. Consider what we might infer about the heights of athletes in general if our sample site was a basketball court as opposed to the jockey's enclosure at a race track. These are obvious biases, but we may not know the biases of our population before we sample them, so we need to take general precautions.

### 4.1.3   How to Select a Random Sample

Let us consider a worked example of how to generate a random sample. For simplicity we will imagine that a small square contains 400 bushes, conveniently arranged at 1 m intervals in 20 rows and 20 columns. Figure 4.1 below shows the weight of fruit (g) from each bush. We need to work out how to select a random sample without knowing anything about the distribution of the population.

How could we do it badly? One of the easiest ways is to walk around and sample the first ten bushes we see. Using this method, we would probably have a bias towards the biggest, most conspicuous, bushes. Our sample of bushes (and probably the yield) would be bigger than the population average, so from our sample we would infer that bushes have a larger yield than they actually do. Also, our bushes would probably all be on the edge, so we may sample bushes that are not competing as much for water and space. This shows how problematic sampling can be, and why time spent planning is not wasted.

To select a position at random we could use the last two digits from the random number button on our calculator. If this gives 0.302 and then 0.420, this would identify the position in row 2 and column 20 (2 m down from the top of the field, and 20 m from the left). If we select a random sample of 16 positions at which to harvest the crop and measure its yield, we could get the selection shown in Figure 4.1. Note that, by chance, no positions have been selected from the top right or bottom right of the field, where crop yields happen to be low. So, this sample is random, but unrepresentative.

| | 1 | 2 | 3 | 4 | 5 | 6 | 7 | 8 | 9 | 10 | 11 | 12 | 13 | 14 | 15 | 16 | 17 | 18 | 19 | 20 |
|---|---|---|---|---|---|---|---|---|---|---|---|---|---|---|---|---|---|---|---|---|
| 1 | 8 | 7 | 12 | 9 | 6 | 8 | 12 | 14 | 7 | 4 | 5 | 4 | 3 | 7 | 4 | 2 | 4 | 6 | 8 | 7 |
| 2 | 16 | 18 | 9 | 10 | 7 | 9 | 14 | 10 | 6 | 3 | 4 | 8 | 5 | 6 | 5 | 3 | 3 | 4 | 6 | 6 |
| 3 | 22 | 15 | 17 | 14 | 12 | 13 | 16 | 9 | 4 | 5 | 6 | 5 | 7 | 7 | 6 | 5 | 4 | 5 | 7 | 9 |
| 4 | 19 | 16 | 13 | 10 | 8 | 9 | 12 | 9 | 8 | 7 | 9 | 10 | 9 | 8 | 3 | 3 | 7 | 10 | 13 | 10 |
| 5 | 16 | 12 | 12 | 9 | 5 | 7 | 10 | 12 | 6 | 8 | 7 | 6 | 6 | 5 | 6 | 4 | 6 | 7 | 9 | 10 |
| 6 | 19 | 16 | 11 | 7 | 8 | 10 | 14 | 10 | 16 | 15 | 17 | 16 | 18 | 17 | 19 | 22 | 24 | 20 | 19 | 15 |
| 7 | 22 | 18 | 16 | 19 | 12 | 19 | 17 | 12 | 18 | 21 | 20 | 19 | 25 | 22 | 24 | 27 | 20 | 17 | 22 | 14 |
| 8 | 25 | 14 | 16 | 12 | 13 | 15 | 16 | 14 | 23 | 25 | 28 | 33 | 30 | 27 | 31 | 25 | 22 | 20 | 24 | 29 |
| 9 | 20 | 17 | 14 | 16 | 17 | 19 | 20 | 24 | 25 | 27 | 32 | 40 | 42 | 35 | 35 | 37 | 29 | 28 | 27 | 35 |
| 10 | 25 | 22 | 19 | 25 | 29 | 32 | 36 | 30 | 35 | 40 | 41 | 47 | 45 | 40 | 32 | 33 | 27 | 27 | 31 | 30 |
| 11 | 29 | 30 | 27 | 32 | 37 | 42 | 45 | 45 | 43 | 51 | 53 | 52 | 48 | 39 | 42 | 37 | 33 | 29 | 27 | 20 |
| 12 | 31 | 35 | 38 | 45 | 47 | 44 | 49 | 51 | 50 | 58 | 56 | 50 | 41 | 43 | 30 | 27 | 29 | 23 | 21 | 19 |
| 13 | 28 | 33 | 30 | 37 | 40 | 39 | 42 | 46 | 51 | 48 | 44 | 40 | 37 | 30 | 35 | 22 | 21 | 17 | 14 | 16 |
| 14 | 26 | 29 | 32 | 35 | 35 | 31 | 29 | 32 | 40 | 37 | 35 | 31 | 16 | 19 | 20 | 11 | 12 | 10 | 13 | 12 |
| 15 | 22 | 28 | 31 | 29 | 28 | 27 | 25 | 27 | 16 | 22 | 19 | 16 | 8 | 4 | 2 | 3 | 6 | 5 | 7 | 10 |
| 16 | 21 | 22 | 28 | 26 | 31 | 27 | 22 | 24 | 20 | 14 | 10 | 8 | 6 | 6 | 4 | 7 | 4 | 3 | 1 | 5 |
| 17 | 17 | 26 | 30 | 31 | 31 | 27 | 25 | 33 | 12 | 8 | 7 | 6 | 6 | 4 | 2 | 1 | 0 | 0 | 0 | 0 |
| 18 | 17 | 24 | 31 | 33 | 27 | 22 | 19 | 17 | 14 | 10 | 9 | 7 | 3 | 0 | 3 | 1 | 0 | 4 | 2 | 0 |
| 19 | 19 | 19 | 21 | 17 | 15 | 16 | 16 | 19 | 16 | 13 | 10 | 12 | 8 | 3 | 1 | 0 | 3 | 2 | 5 | 3 |
| 20 | 23 | 21 | 17 | 14 | 13 | 19 | 23 | 17 | 12 | 11 | 6 | 9 | 7 | 3 | 1 | 0 | 1 | 1 | 2 | 4 |

**FIGURE 4.1**
Random sample of 16 bushes. Bold numbers indicate samples.

## 4.1.4 Systematic Sampling

In systematic sampling, sample units are chosen to achieve maximum dispersion over the population. They are not chosen at random but regularly spaced in the form of a grid (Figure 4.2).

The first point of a systematic sampling grid should be chosen at random so that we do not generate a systematic bias, for example, always miss out the edges of a site. Once this is chosen (for example, column 15, row 2 in Figure 4.2), the other sample points are chosen in a fixed pattern from this point, and so they are not independent of each other.

Much use is made of systematic sampling. For example, every 10th tree in every 10th row may be measured in forestry plantations. As long as the number of sample units is high, there is little risk of coinciding with any environmental pattern that might affect tree growth (e.g., drainage channels), and the data are often treated as if they were from a random sample.

| | 1 | 2 | 3 | 4 | 5 | 6 | 7 | 8 | 9 | 10 | 11 | 12 | 13 | 14 | 15 | 16 | 17 | 18 | 19 | 20 |
|---|---|---|---|---|---|---|---|---|---|---|---|---|---|---|---|---|---|---|---|---|
| 1 | 8 | 7 | 12 | 9 | 6 | 8 | 12 | 14 | 7 | 4 | 5 | 4 | 3 | 7 | 4 | 2 | 4 | 6 | 8 | 7 |
| 2 | 16 | 18 | 9 | 10 | 7 | 9 | 14 | 10 | 6 | 3 | 4 | 8 | 5 | 6 | 5 | 3 | 3 | 4 | 6 | 6 |
| 3 | 22 | 15 | 17 | 14 | 12 | 13 | 16 | 9 | 4 | 5 | 6 | 5 | 7 | 7 | 6 | 5 | 4 | 5 | 7 | 9 |
| 4 | 19 | 16 | 13 | 10 | 8 | 9 | 12 | 9 | 8 | 7 | 9 | 10 | 9 | 8 | 3 | 3 | 7 | 10 | 13 | 10 |
| 5 | 16 | 12 | 12 | 9 | 5 | 7 | 10 | 12 | 6 | 8 | 7 | 6 | 6 | 5 | 6 | 4 | 6 | 7 | 9 | 10 |
| 6 | 19 | 16 | 11 | 7 | 8 | 10 | 14 | 10 | 16 | 15 | 17 | 16 | 18 | 17 | 19 | 22 | 24 | 20 | 19 | 15 |
| 7 | 22 | 18 | 16 | 19 | 12 | 19 | 17 | 12 | 18 | 21 | 20 | 19 | 25 | 22 | 24 | 27 | 20 | 17 | 22 | 14 |
| 8 | 25 | 14 | 16 | 12 | 13 | 15 | 16 | 14 | 23 | 25 | 28 | 33 | 30 | 27 | 31 | 25 | 22 | 20 | 24 | 29 |
| 9 | 20 | 17 | 14 | 16 | 17 | 19 | 20 | 24 | 25 | 27 | 32 | 40 | 42 | 35 | 35 | 37 | 29 | 28 | 27 | 35 |
| 10 | 25 | 22 | 19 | 25 | 29 | 32 | 36 | 30 | 35 | 40 | 41 | 47 | 45 | 40 | 32 | 33 | 27 | 27 | 31 | 30 |
| 11 | 29 | 30 | 27 | 32 | 37 | 42 | 45 | 45 | 43 | 51 | 53 | 52 | 48 | 39 | 42 | 37 | 33 | 29 | 27 | 20 |
| 12 | 31 | 35 | 38 | 45 | 47 | 44 | 49 | 51 | 50 | 58 | 56 | 50 | 41 | 43 | 30 | 27 | 29 | 23 | 21 | 19 |
| 13 | 28 | 33 | 30 | 37 | 40 | 39 | 42 | 46 | 51 | 48 | 44 | 40 | 37 | 30 | 35 | 22 | 21 | 17 | 14 | 16 |
| 14 | 26 | 29 | 32 | 35 | 35 | 31 | 29 | 32 | 40 | 37 | 35 | 31 | 16 | 19 | 20 | 11 | 12 | 10 | 13 | 12 |
| 15 | 22 | 28 | 31 | 29 | 28 | 27 | 25 | 27 | 16 | 22 | 19 | 16 | 8 | 4 | 2 | 3 | 6 | 5 | 7 | 10 |
| 16 | 21 | 22 | 28 | 26 | 31 | 27 | 22 | 24 | 20 | 14 | 10 | 8 | 6 | 6 | 4 | 7 | 4 | 3 | 1 | 5 |
| 17 | 17 | 26 | 30 | 31 | 31 | 27 | 25 | 33 | 12 | 8 | 7 | 6 | 6 | 4 | 2 | 1 | 0 | 0 | 0 | 0 |
| 18 | 17 | 24 | 31 | 33 | 27 | 22 | 19 | 17 | 14 | 10 | 9 | 7 | 3 | 0 | 3 | 1 | 0 | 4 | 2 | 0 |
| 19 | 19 | 19 | 21 | 17 | 15 | 16 | 16 | 19 | 16 | 13 | 10 | 12 | 8 | 3 | 1 | 0 | 3 | 2 | 5 | 3 |
| 20 | 23 | 21 | 17 | 14 | 13 | 19 | 23 | 17 | 12 | 11 | 6 | 9 | 7 | 3 | 1 | 0 | 1 | 1 | 2 | 4 |

**FIGURE 4.2**
Systematic sample of 16 bushes.

In social surveys, every 50th name on the electoral roll might be selected as a person to be interviewed. This is very convenient. However, it is important to be aware of the possibility of bias. If flats were in blocks of 25 and were all occupied by couples, we might only interview people who lived on the ground floor!

Systematic sampling is very efficient for detecting rare events because we are less likely to miss one such event (perhaps a fallen tree within a wood or a molehill in grassland) than if the sampling has a random element. Also, because we are more likely to include both very small and very large individuals, the mean of a homogeneous population is often close to the true mean but has a large standard error.

## 4.2 Comparing More Than Two Groups

So far we have learned how to compare means of two groups. Very often, however, we wish to compare the mean performance of several different categories. For example:

- Do four fertilisers differ in their crop yields?
- Do three species of plant differ in their ability to colonise loam?
- Do five drugs differ in their ability to cure disease?

In this section we will see how to design experiments that will answer such questions. Let us start by designing an experiment to investigate the effect of fertiliser on yield of barley. We will have three treatments (A, B, and C). The term treatment is used to describe what the crop receives — so we are going to design an experiment with three treatments.

## 4.3  Principles of Experimental Design

### 4.3.1  Objectives

It is good practice to write down the background to your experiment. This consists of why you are interested in the problem and your general objective. It is good to include any important aspects of the biology of your experimental organism, or any practical considerations that may affect the experimental layout. Do you need to leave space between plots to avoid cross-contamination? Do you need to leave access space for watering or feeding? It is best to identify any biological factors that will affect your layout early on.

### 4.3.2  Replication

Replication gives us an indication of the variation of results and, in this way, an idea of the accuracy of our estimates. Suppose you are a policeman attending an accident. If you arrive on the scene and five separate witnesses tell you that the car was blue, you would be much more confident that the car was blue than if there were only one witness. Similarly, suppose four witnesses say blue, and one says green. You would still probably conclude that the car was blue, but with less certainty than before. Replication has also given you an idea of the variation. If you had spoken to the first witness (green) and stopped there, you would have concluded the wrong color and have no idea about the spread of opinion.

This concept is illustrated graphically below (Figure 4.3). Note how increasing the number of replicates dramatically changes the relationship and the inferences you would draw.

So, how do we replicate each treatment in a planned experiment (Figure 4.4)?

It is no good just splitting a field in three and applying a different fertiliser to each third (Figure 4.4a). Perhaps the natural fertility of the soil is higher at the bottom of the slope, so whichever fertiliser is allocated to that position will appear to be even better than it really is in comparison with the others. There is only one replicate of each treatment so how can we tell whether the

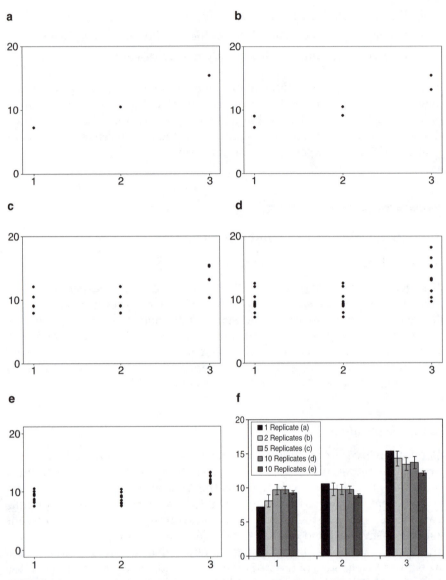

**FIGURE 4.3**

(a–d) As the number of replicates increases, the shape of the relationship changes. We can have more confidence in the estimate with more replication. (e): A data set with the same group means but smaller spread. (f) The relationship between replication and the standard error. In each case, the standard error decreases with increasing replication, as shown by the error bars getting smaller. Standard error cannot be calculated for only one group, so the error bar is missing for this data series. As shown by the difference between (e) and (f), less spread gives a smaller standard error, which indicates that we have a better estimation of the mean.

effect is due to the treatment, or to soil differences between plots? However, if we divide the field into 12 parts or experimental units (usually called plots) and allocate each fertiliser treatment (Section 4.3.3 discusses how to randomly

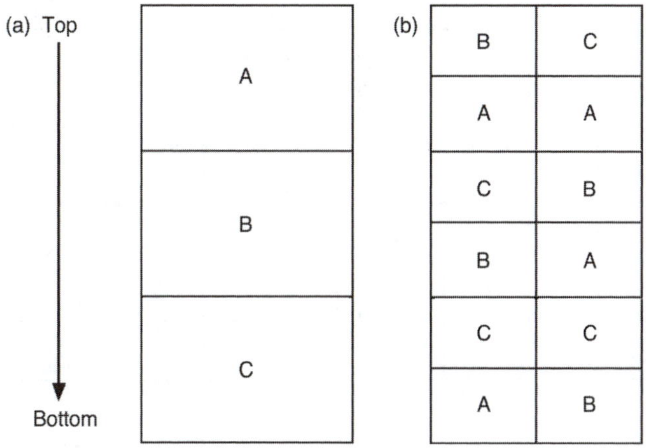

**FIGURE 4.4**
(a) Allocating treatments without replication. (b) An experiment with four replicates, each randomly allocated to a position on the field.

allocate treatments) to four of them, we will improve matters (Figure 4.4b). We now have four replicates of each treatment.

We can compare the yields from the four plots receiving organic fertiliser treatments. Any variation in these yields is caused by random variation in conditions across the site. The same is true for variation in the four yields from treatment B and, separately, from treatment C. So, replication has given us three separate estimates of the background or random variation in the yields of plots receiving the same treatment.

The greater the replication we have (perhaps six plots of each treatment instead of four), the more independent pieces of evidence we have about the fertilisers' effects. This means that we are able to detect smaller real differences in yield between the populations from which we have taken our samples. We can achieve increased precision in our estimates of the population mean yields and of differences between them by having more replication. A more technical discussion on choosing how much replication is needed to be found in Appendix B.

### 4.3.3 Randomisation

The four replicate plots of each treatment must be allocated to positions in the field at random to avoid the unknown biases we discussed before (Section 4.1.3). This is achieved by numbering the 12 plots from 1 to 12 (Figure 4.5a). We then use the random number button on the calculator (or use random number tables) to select 4 of these plots for treatment A (e.g., the numbers 0.809, 0.312, 0.707, 0.836, and 0.101 allocate this treatment to plot numbers 9, 12, 7, and 1; we ignore the value 36 from 0.836 because it is greater than 12) (Figure 4.5b). Then we select four more numbers for treatment B, and treatment C must go on the four plots that remain, which will be random because we randomised A and B.

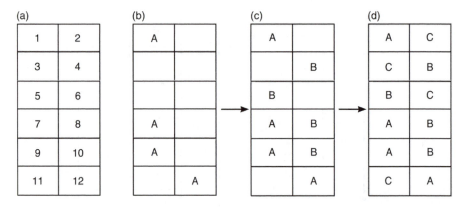

**FIGURE 4.5**
(a) Numbering of plots to enable random allocation of treatments (b–d) Allocating each treatment to a plot in turn.

As with selecting a sample from one population (Chapter 2), we can think of randomisation in allocating treatments to experimental units as an insurance policy. It protects us from obtaining estimates of treatment means that are biased, that is, consistently higher or lower than the population mean. This could arise because all the plots for one treatment happened to be in the corner of the field where the crop was damaged by frost. The replicates of each treatment must be interspersed (mixed up) over the site, and randomisation is a good way of achieving this. Most statistical analyses assume that you have randomised experimental units to the treatments. Very rarely, randomisation may produce an arrangement of plots in which, say, the four replicate plots of one treatment are grouped together at the bottom of the slope. Statistically, this would be fine because randomisation works to ensure that experiments in general are bias free. However, we are only interested in this particular instance of the experiment, which will cost us time and money. In the case of randomisation producing a clear bias, it would be prudent to rerun the randomisation process and reallocate plots. Only do this in very clear cases, however. Humans are very good at detecting patterns, and as a consequence, we tend to think we see clusters in patterns that are actually truly random.

### 4.3.4 Controls

Should we have a control in an experiment? A control is the name given to a treatment in which nothing is applied to the plot. We can then see what changes take place naturally during the experiment. Statistically, we treat this as an extra treatment. A variation on this idea is a procedural control. For example, in a drug trial, we might inject three different drugs into people (the treatments), and the control would be injecting saline. This means that we are not comparing people who have had an injection with people who have not.

So far we have been considering a simple experiment in which fertilisers are applied to a field. Imagine a more complicated experiment in which we are interested in the effect of sheep grazing at different times of year on the number of wildflower species in grassland. We can fence off each plot to make sure that the sheep are in the right place at the right time each year for, say, 5 years. However, the grassland present at the beginning will contain a range of species characteristic of its location, soil type, and previous management (for example, always grazed with sheep or was growing maize 2 years ago). So, if we carried out the same experiment on another site 50 km away, the species present would be rather different. Thus, having control plots on each site is useful. It tells us what happens on each site in the absence of any grazing treatments. We can think of it as providing a standard comparison between the two sites.

### 4.3.5 Blocking

What we are doing in experimental design is making it easy to partition variation into what we are interested in and what we are not. We will cover this in greater detail in Chapter 5, but in the meanwhile, it is important to introduce the concept of **blocking**. In Section 4.3.3 we looked at randomisation as a way of avoiding experimental biases. Another way to do this is by blocking, and the two techniques are routinely used together. Whereas randomisation is used to control for *unknown* biases or variation, blocking is brought in to control for *known* or *likely* differences between replicates.

With the completely randomised distribution of treatments earlier in this chapter, it is quite likely that we will allocate two replicates of a particular treatment to the field margin at the top of the slope and none to the margin at the bottom. Therefore, our results for each fertiliser could be influenced as much by the sample location as the effect of fertiliser. If we were to carry out such an experiment many times, effects would even out because, perhaps next time, the same treatment might be overrepresented at the bottom of the slope. However, if we have only the resources for one or two experiments, we need to find a way of overcoming known variations like this. Look at the two following graphs; see how our interpretation changes if we know that the higher values in each treatment come from wetter soil. If we can find a way of including this information in the analysis, then the error (the spread of the results) will clearly be much less (Figure 4.6).

If we know that the soil is dry at the top of our field and wet at the bottom, then we should block for this rather than randomise. We have four replicates, so it makes sense to divide the field into four blocks and represent each of our three treatments in each block. We should still randomise the position of treatments within block, but the main (known) bias is already dealt with by block. The logic of blocking is that rather than averaging out a bias, we are investigating it like a treatment, so that we work out the effect it has and then discount it. How we analyse a design that includes a blocking factor will be covered in Chapters 5 and 6.

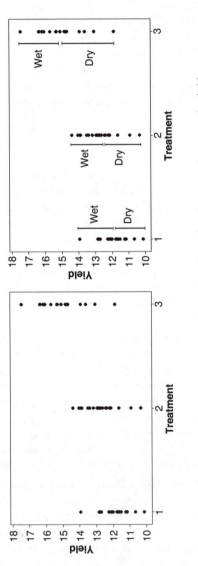

**FIGURE 4.6**
How our understanding may be further improved by including the information that soil moisture varies across the field.

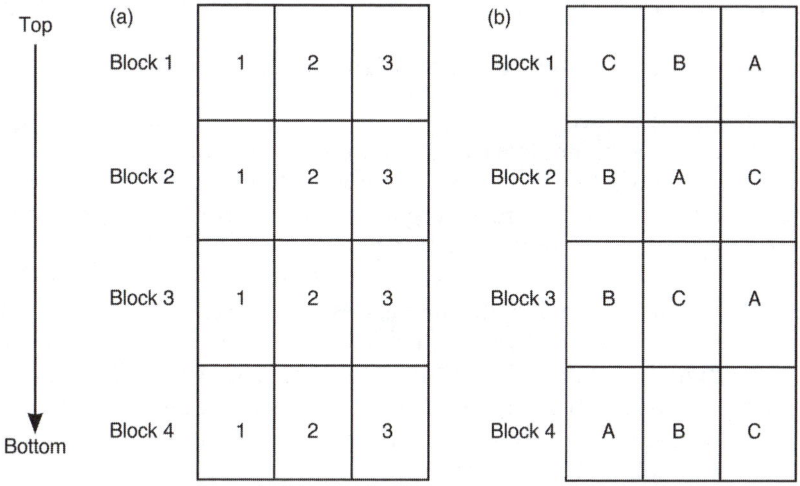

**FIGURE 4.7**
A randomised complete block design; each treatment is present in each block, but the order within each block is randomised.

A **randomised complete block design** (Figure 4.7) and reduces random variation and improves our ability to detect differences between treatments. Each quarter of the experiment is a "block," and each block contains a complete set of all the treatments within it. Each block is selected so that the conditions are even (or homogeneous) *within* it but they differ *between* one block and another. So, if the preceding diagram represents a field the top block might be on slightly sandier soil, and the block at the bottom might be more shaded because of a hedge. These differences should not be too great, as blocking is just a way of getting rid of confusing variation and will not give reasonable results unless the whole experiment is conducted on a reasonably uniform site.

How should we allocate the four treatments to the three plots within each block? This must be done using random numbers (as in Figure 4.7), but here we would generate random numbers from 1 to 3 to ensure that each treatment has an equal chance of occurring on each plot. We number the four plots in each block from 1 to 3. Then we use the random number button on our calculator. If the last digit is 2, we allocate treatment A to plot 2; if number 1 appears next, we allocate treatment B to plot 1. This leaves plot 3 for treatment C. Repeat this process for the next block.

Blocking should be used wherever there may be a trend in the environment that could affect the feature in which you are interested. For example, in a glasshouse heating, pipes may be at the rear of the bench, so one block should be at the rear and another at the front of the bench. Even in laboratories and growth cabinets there can be important gradients in environmental variables that make it worthwhile arranging your plots (pots, trays, petri dishes, etc.) into blocks. It is common to block feeding experiments with animals by putting the heaviest animals in one block and the lightest ones in another block. This helps to take account of differences in weight at the start of the experiment.

## 4.4  Recording Data and Simulating an Experiment

Considering how much time and resources you will spend on your experiment, it should be compulsory to simulate it before you start. Think about what you are going to measure and how you should lay it out in your lab or field notebook. If you are really organised, it is worth creating a data table and printing out copies of it to stick in your notebook. Try to copy the data onto a spreadsheet as soon as possible. Aside from providing a copy, this will highlight mistakes while the data is fresh in your mind. Photocopy data sheets whenever you can; if you cannot do this in the field, a useful substitute is to take digital photos at the end of every day. These are usually of sufficiently high quality to be able to recreate your data if it gets lost. In the real world, pages get wet or torn out, so do not forget to number every page or subject (worm, plant, bacterial plate).

A huge number of undergraduate projects cannot be analysed at all or at least in the way that was anticipated. This may be due to design, or loss of data points. Data points are lost by plants or animals dying, experimental plots getting disturbed, or by someone clearing out the fridge in the lab. Murphy's Law is alive and well in most experiments. Although in principle we should be able to get the design correct, it is a good idea to create dummy data using random numbers, and use these to try out your proposed analysis. If the computer does not let you run the analysis you want, it may be that you have insufficient replication or some other problem that you could fix before starting the experiment. If this analysis works, try deleting randomly selected data points to see how robust your design is. If the loss of one or two data points prevents you from running the test you want, you might want to reconsider the design. In addition to the factor of interest, take note of unusual observations that might influence the results. You might be able to account for them later, or it could lead onto another study.

The take-home message is that time spent planning is never wasted. Discussing your experimental design with an experienced researcher will identify most pitfalls and probably throw up more ideas to incorporate. Having a clear idea of what you are doing and exactly what to record saves a lot of time in the field or lab. *If nothing else, carry out a dummy analysis to see if you can analyse your experiment according to your plans.*

## 4.5  Simulating Your Experiment

You can create dummy data and analyse them for very complicated experiments. However, we will show you how to do this with a simple experimental design which can be analysed with a test we have already covered (the two-sample t-test).

Suppose you have been asked to carry out an experiment to compare the effects of two different insect diets on the amount of frass (insect feces) produced. You have 18 insects available, and you can choose which diet each insect receives. The data collected is the dry weight of frass produced by each insect.

### 4.5.1    Creating the Data Set

We have 18 sites and 2 treatments, so we need the data in Table 4.1.

You can type this in, or get MINITAB to do it for you. Label the columns, and then go to **Calc>Make Patterned Data> Simple Set of Numbers> Store patterned data in:** Insect **From first value:** 1 **To last value:** 18 **In steps of:** 1 **List each value:** 1 **List whole sequence:** 1

Do this again for Treatment, using first value 1, last value 2, and listing each value once, and you should get the data provided in Table 4.1. This is fine, but we clearly have not randomly allocated the sites to each treatment. We can do this by making use of a trick in MINITAB: **Calc>Random Data>Sample from Column(s)>Sample** 18 **rows from column(s):** Treatment **Store samples in:** Treatment. Make sure that the sample with replacement box is **not** checked. This randomises treatment with respect to insect number, and you should get something like Table 4.2. Note that if you do this yourself, you should get something similar, but not exactly the same because of the randomisation process.

**TABLE 4.1**

Insect Number and Treatment for Insect Experiment

| Insect | Treatment |
| --- | --- |
| 1 | 1 |
| 2 | 1 |
| 3 | 1 |
| 4 | 1 |
| 5 | 1 |
| 6 | 1 |
| 7 | 1 |
| 8 | 1 |
| 9 | 1 |
| 10 | 2 |
| 11 | 2 |
| 12 | 2 |
| 13 | 2 |
| 14 | 2 |
| 15 | 2 |
| 16 | 2 |
| 17 | 2 |
| 18 | 2 |

**TABLE 4.2**

Treatments Randomised to Insects

| Insect | Treatment |
|--------|-----------|
| 1 | 1 |
| 2 | 2 |
| 3 | 1 |
| 4 | 2 |
| 5 | 2 |
| 6 | 1 |
| 7 | 1 |
| 8 | 2 |
| 9 | 1 |
| 10 | 2 |
| 11 | 1 |
| 12 | 1 |
| 13 | 2 |
| 14 | 2 |
| 15 | 2 |
| 16 | 2 |
| 17 | 2 |
| 18 | 1 |

**TABLE 4.3**

Simulated Data for the Insect Experiment

| Insect | Treatment | Frass Weight |
|--------|-----------|--------------|
| 1 | 1 | 12.9013 |
| 2 | 2 | 12.6835 |
| 3 | 1 | 10.2610 |
| 4 | 2 | 9.7528 |
| 5 | 2 | 11.3427 |
| 6 | 1 | 14.1971 |
| 7 | 1 | 11.6366 |
| 8 | 2 | 13.8637 |
| 9 | 1 | 12.5817 |
| 10 | 2 | 13.8720 |
| 11 | 1 | 11.4658 |
| 12 | 1 | 11.4404 |
| 13 | 2 | 11.1036 |
| 14 | 2 | 10.9567 |
| 15 | 2 | 12.9425 |
| 16 | 2 | 10.3909 |
| 17 | 2 | 12.4214 |
| 18 | 1 | 12.4716 |

Finally, you need to simulate the data (the things you will record in the experiment). Do this by: **Calc>Random Data>Normal>Generate** 18 **rows of data Store in column(s):** Frass weight **Mean:** 12 **Standard deviation:** 1.0.

If you already have an estimate of the mean and standard deviation of whatever you are measuring, include it for realism, but it does not matter too much.

The simulated data are shown in Table 4.3.

## 4.5.2  Analysing the Simulated Data

Carry out a two-sample t-test as described in Chapter 3, and you should get something similar to the following output. The two groups are not significantly different, but this is hardly surprising, as we have just simulated the data using the same mean for each group. What this does tell us is that we can analyse the data in the way we intended and that there are no nasty hidden surprises.

```
Two-Sample t-Test and CI: Frass Weight, Treatment
Two-sample t for Frass weight

Treatment  N   Mean   StDev  SE Mean
        1  9  12.08   1.29     0.43
        2  9  11.96   1.38     0.46


Difference = mu (1)   mu (2)
Estimate for difference: 0.120598
95% CI for difference: (1.217092, 1.458289)
t-Test of difference = 0 (vs. not =): T-value = 0.19
P-value = 0.850 DF = 15
```

# 5

## Partitioning Variation and Constructing a Model

"I checked it very thoroughly," said the computer, "and that quite definitely is the answer. I think the problem, to be quite honest with you, is that you've never actually known what the question is."

**Deep Thought, in *The Hitchhiker's Guide to the Galaxy*, by Douglas Adams**

## 5.1  It's Simple . . .

So far in this book we have tried to make clear the underlying reasoning behind the analysis of simple experimental and observational data. The basic notion is that most hypothesis tests concern the distribution of test statistics, which are quantities we can estimate from our data and which have a known distribution if the Null hypothesis is true. By careful choice of the test statistic, we can find out whether sets of measurements conform to expectations (Chapter 2), or whether groups of organisms produce different results when they are subjected to different treatments (we have seen how to compare two groups in Chapter 3). In addition to the factors we control in our experiments, there are many other influences that may affect the results (as discussed in Chapter 4). Some extraneous influences can be counteracted by using appropriate sampling methods, but others must be taken into account in the statistical analysis. This chapter deals with how we analyse the experimental designs described in Chapter 4.

## 5.2  . . . But Not That Simple

Before R. A. Fisher developed the modern approach to statistics, scientists held the view that to work out the effect of one treatment, you should vary it experimentally while holding all other variables constant.

This suffers from two drawbacks; first, the world is complicated, and we might not be able to strictly control everything, especially if our experiment is in the field. Second, investigating only one thing at a time can lead to the wrong answer (or a very incomplete one). Think of a car with one gear. Its speed is determined by how far the accelerator and brake pedals are pressed down. If you only measured the effect of the accelerator on the car's speed, you would not have a clear result. This is because the effect of the accelerator on the speed also depends on action of the brake. Stepping on the accelerator has no effect on the speed when the brake is fully on. So, *the effect of one factor can depend on the action of another.* As Deep Thought suggested, we may not have a complete answer because we have not asked the complete question. Investigating the equivalent of only one pedal at a time can be very misleading. A good real-world example is of one very effective clinical drug in use today that can have side effects of nausea and vomiting. Whether or not you suffer this side effect depends on whether or not you eat pineapple with it. You can be fairly sure that the effect of pineapple was not investigated in the original drug trials. Fisher showed that you can get a much more complete answer by deliberately making the experiment more complicated and investigating more than one factor at the same time.

## 5.3 The Example: Field Margins in Conservation

As part of a government environmental scheme, farmers are being paid to sow grass borders around the edge of fields. There is pressure to add wildflower seed into the sowing mix, but at additional cost. It is thought that grass borders will increase the local invertebrate populations, with consequent beneficial effects on biodiversity in general (Baines et al., 1988). As an indicator, we have decided to count spider numbers. Will the wild flower seed increase the number of spider numbers per plot?

How many times per year the margins should be cut is also of interest, and the effects may be different in the grass-only treatment from the grass–wild flower mixture. We will therefore investigate these two factors in one experiment. This is called a **factorial** design, and if we investigate all the combinations, it looks like this (Table 5.1):

**TABLE 5.1**

Factor Levels in Different Treatment Combinations

| | Factor 1: Cutting | |
|---|---|---|
| **Factor 2: Seeding** | **Level 1: Cutting Once/Year** | **Level 2: Cutting Twice/Year** |
| Level 1: with flower seeds | Treatment 1: F1 | Treatment 2: F2 |
| Level 2: no flower seeds | Treatment 3: NF1 | Treatment 4: NF2 |

## 5.4   The Idea of a Statistical Model

So far we have managed to get by without a very formal understanding of scientific explanation involving statistical analysis, but as things become more complicated, we need to clarify the relationship between what we are trying to explain and how we will do it. The basic idea is that we can represent our understanding in the form of an equation, with the left-hand side representing what we are trying to explain, and the right-hand side the things we think explain it. Thus, a statistical model, for the example, could take the form

$$spidernumbers = treatment + error$$

This expresses the idea that the number of spiders in a quadrat depends on which treatment combination was applied and on all the other things that affect spiders but which we have not been able to include in the model. This second component is called "error" because, in the original formulation of his ideas, Fisher thought that it would be due mainly to measurement errors, but we now use the term to describe variation not explained by experimental treatment. In fact, when stating a model formula, the error term is not usually included because it is assumed to be present in all models. The usual formulation for the example involving sheep growth on different diets (Section 3.5) would therefore be

$$growth = diet$$

It is important to realise that these are not mathematical equations, although they may look like them. Clearly, spider numbers (a whole number) cannot be "equal to" treatment, which is a list of labels describing what we did to the various quadrats. We should understand "=" as meaning "is explained by." In fact, to be strictly accurate, we are saying that the *variation* between plots in their spider numbers is explained by the treatment. For simple cases such as these, the model formula does not really add much to our understanding, but consider the following:

$$spidernumbers = cuttingregime + seedsowing$$

This model proposes that spider numbers depend on both the cutting regime and on whether seeds are sown. Furthermore, the use of the "+" between the terms represents the proposal that these two effects add together in their effects on spider numbers; this implies that they act independently of one another. To represent the proposal that the action of one might depend on the other (as in the accelerator and brake example in the previous section), we would have to write:

$$spidernumbers = cuttingregime + seedsowing + cuttingregime*seedsowing$$

where the extra term *cuttingregime\*seedsowing* represents the extent to which the two factors **interact** with one another.

We will have more to say about statistical model formulae, but for now a final point is to see how the formula relates to the hypothesis tests involved. Each term in the model represents a separate hypothesis about how the term on the left-hand side (which we call the **response variable**) is affected by the things we think might be affecting it (which are often called **predictors** or **predictor variables**). Thus the model

$$spidernumbers = treatment$$

represents the research hypothesis that at least one of the treatments will have a different mean number of spiders from the others. The Null hypothesis is therefore that all treatment plots have similar means, and any differences are due to random noise in the system. The terms in the model

$$spidernumbers = cuttingregime + seedsowing + cuttingregime*seedsowing$$

represent three different hypotheses, namely

- cutting affects spider numbers
- seed sowing affects spider numbers
- the effect of cutting depends on seed sowing and vice versa

## 5.5  Laying Out the Experiment

Whether or not we can test the hypotheses represented by the model depends on getting the experimental design right. In the last chapter, we discussed replication, randomisation, blocking, and practical considerations, so we will consider each of these in turn.

### 5.5.1  Replication

So, how many replicates? The simple answer is as many as you can, but be realistic! Too many replicates are inefficient, but the more replicates we have, the more precision we will have in our results. On the other hand, we will have more work to do! We need to consider the cost and availability of resources: people, land, equipment, and the time and effort involved. Tired people will start to make mistakes at the end of a hard day's work — no matter how dedicated they are. Somewhere there is a compromise to be reached. If you are the sort of person that can work 12 h a day for an entire field season, you might consider 82 replicates. If you are not, do not. How many fields/

beetles/plants can you actually afford or have access to? Obviously you must have enough data to have some hope of answering the question. You do not want to end up making a Type II error (accepting $H_0$ when it is false: Section 2.6) just because you did not collect enough data. A more technical discussion of this point is found in Appendix B. It is important to be realistic because, if you cannot get enough replicates for a good analysis, *walk away*. A poor experiment that does not answer the question achieves nothing. Leave it to someone with more time or money rather than do the work and then find out that you do not have enough replicates to detect the signal from the background noise.

### 5.5.2 Randomisation and Blocking

Although randomisation is a more generally applicable (though less powerful) method of controlling unwanted variation in an experiment (hence it was dealt with first in Chapter 4) we have to decide about the blocking structure before we can randomly assign plots to treatments. Because each treatment must be present in each block, we randomise within blocks as in Figure 5.1 to ensure against any hidden biases within blocks.

There are four replicates of each treatment in our experiment, and the 16 plots required are located at random with four plots on each side of the field. However, the four sides of the field may well provide slightly different environments. This

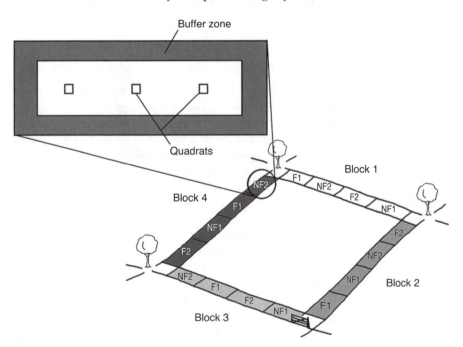

**FIGURE 5.1**
Randomised complete block showing practical elements of experimental design, such as a buffer zone and quadrats.

field may be on a slope; the side at the top may have drier or sandier soil than that at the bottom; perhaps a tall hedge runs along one side of the field, whereas there may be only a low fence along the other side. With the completely randomised distribution of treatments, it is very likely that we will have allocated two replicates of a particular treatment to the field margin at the top of the slope and none to the margin at the bottom. Such differences can be important because, for example, in a dry season, grass growth will be less vigorous at the top of the slope, and this may mean that low-growing wild-flowers have a better chance to get established. Thus, if our results from the treatment in which we sowed flower seeds and which we cut once (F1) show it to have a large number of spiders, this may partly be because of its over-representation on the favourable ground at the top of the slope. If we were to carry out such an experiment many times, such effects would even out because, perhaps next time, the same treatment might be overrepresented at the bottom of the slope. However, if we have only the resources for one or two experiments, we need to find a way of overcoming variations like this.

Ideally, we want one replicate of each treatment to be on each of the four sides of the field in a randomised complete block design as in Chapter 4. Each side of the field is a block, and each block contains a complete set of all the treatments within it.

Now we follow the procedure shown in Chapter 4 and number the plots within each treatment. Use a random number table or the random function on your calculator to allocate the treatments.

### 5.5.3   Practical Considerations

We have decided to have four replicate lengths of strip (plots) for each of the treatments. We must decide the size of each strip. Its depth from the edge of the field to the crop will be as used by farmers (say, 2 m?), but how long should a plot be? It is important to have enough room to turn mowers around at the ends of each plot. Also, it is possible that wildflower seeds may blow a meter or so into a neighbouring plot during sowing.

When a particular plot has received its one cut of the year, it will grow quite tall in the rest of the year and may shade the edge of the neighbouring plot that receives a further cut. All these facts point to the idea of having a buffer zone round the edge of each plot. This is treated as far as possible like the rest of the plot, but we will not make any recordings from it because it is unlikely to be a good representation of the treatment. Its function is to protect the inside of the plot.

We can now record what is going on by marking the position of small areas within each plot (or quadrat) from which we will take our samples. It is common to use say a $0.5\,\text{m} \times 0.5\,\text{m}$ or a $1\,\text{m} \times 1\,\text{m}$ quadrat for recording vegetation. There will be variation within each plot — perhaps caused by soil composition, shading by occasional large shrubs in the nearby hedge, or rabbit grazing. So, we should sample several quadrats within each plot.

We can then take the mean value from them as a fairer representation of the plot as a whole than if we had recorded only one quadrat that happened by chance to have been on the regular pathway of a badger. Let us assume that we decide to have three quadrats on each plot. This will allow us to decide on the length of the plot after allowing room between each quadrat to move around and place equipment. We now have $16 \times 3 = 48$ quadrats to record in total.

---

## 5.6   Sources of Variation: Random Variation

It is important to realise that, even if we did not impose any treatments in the experiments, we would find that the number of spiders of a particular species would not be the same in each quadrat. As discussed above and in Chapter 4, we can minimise these effects by randomizing and blocking, but a certain amount of variation will always be present. Such natural variability from one quadrat or sampling unit to another is called **random variation** and is represented in the model formula by the (implied) error term.

In a laboratory experiment, such variation is often quite small. For example, the effect of different concentrations of a chemical on the growth rate of a bacterium may be studied by adding the chemical solutions to inoculated agar in petri dishes that are then kept at a constant temperature. The conditions within each petri dish will be extremely similar, apart from the deliberately imposed treatments (the chemical concentrations) being different. However, they will still vary to some extent, and this random variation should be taken into account. The model formula approach expresses clearly the idea that we must separate background or random variation from that caused by the treatments.

Let us look at the results we might collect from our field experiment. First, we count the number of spiders in each quadrat in our plot. We then add together the values from each of the three quadrats per plot and divide by three to give the number of spiders per meter square. This is because it is the plots that were randomised, and so they are the experimental units. Recording three quadrats within each plot is like subsampling the plots. It gives us more information and allows us to obtain a better estimate of the number of spiders per square meter of plot than if we had only recorded one quadrat. However, because each set of three quadrats is within one plot, the quadrats are not independent observations and therefore should not be used separately in the analysis of variance that is described in the following text. Thus we can summarise the variation in spider numbers across the 16 plots as shown in Table 5.2.

We find that the mean number of spiders on plots that received wildflower seed and were cut once (F1) was 19.5, whereas, on those that were not sown and were cut twice a year (NF2), the mean was only 13.0. Not every plot receiving treatment F1, however, will contain 19.5 spiders (this is simply an average — we cannot have half a spider in reality). There is a considerable

**TABLE 5.2**

Number of Spiders per Plot

| | Treatment | | | |
|---|---|---|---|---|
| Replicate | F1 | F2 | NF1 | NF2 |
| 1 | 21 | 16 | 18 | 14 |
| 2 | 20 | 16 | 17 | 13 |
| 3 | 19 | 14 | 15 | 13 |
| 4 | 18 | 14 | 16 | 12 |
| Mean | 19.5 | 15.0 | 16.5 | 13.0 |

amount of random variation around the treatment mean. This variation is the unexplained or "error" variation.

## 5.7   The Model

We can set up a mathematical model that describes these ideas. This is a system of equations we can use to work out the expected values. Although it looks a bit similar to (and is related to), the "model formula" of Section 5.4, here we are dealing with actual mathematical equations. If we want to predict the number of spiders on a plot, the simplest estimate would be the mean of all 16 values. This is called the **grand mean**. It is 16.0.

We can do better than this. We observe that if a plot is in treatment F1, its mean number of spiders will be greater than the grand mean. The difference between the treatment mean and the grand mean, it is $19.5 - 16.0 = +3.5$. So the best estimate would be that an F1 plot has 3.5 more spiders than the grand mean. In contrast, for treatment NF2 this difference is $13.0 - 16.0 = -3.0$. So on average NF2 plots have 3.0 spiders *less* than the grand mean.

We can represent this more generally as

$$Expected\_number = grandmean + (treatmentmean - grandmean)$$

This simple model predicts that each of the four plots in treatment F2 is expected to have $16 + (15 - 16) = 16 + (-1) = 15$ spiders per square meter. However, they do not! So, the model needs further work.

We can find out by how much our model fails to fit our observations on each plot in treatment F2 by subtracting the expected value from the observed value (Table 5.3). Two of the replicate plots of this treatment have observed numbers greater than the mean for the treatment and two have values that are less. The differences between observed and expected values are called **residuals**. They represent the random or *error* variation. Residuals can be positive or negative, but they always add up to zero for each treatment and so they must also have a mean of zero.

**TABLE 5.3**

Example of Observed Values, Expected Values, and Residuals for Treatment F2

| Replicate | Observed Number | Expected Number | Difference (Residual) |
|-----------|-----------------|-----------------|-----------------------|
| 1 | 16 | 15 | +1 |
| 2 | 16 | 15 | +1 |
| 3 | 14 | 15 | −1 |
| 4 | 14 | 15 | −1 |
| Mean | 15 | 15 | 0 |

So, a simple mathematical model to explain what is going on in our experiment is

*Observed number of spiders per plot = expected number of spiders per plot + residual.*

Or, in more general terms:

*Observed value = expected value + residual.*

Because we already know how to calculate an expected value (see previous text), we can include this information as well to give the full equation:

*Observed_number = grandmean + (treatmentmean − grandmean) + residual*

We can make this clearer by using the term **treatment effect** to represent the difference between the treatment mean and the grand mean.

*Observed_number = grandmean + treatment_effect + residual*

Note that both the treatment effect and the residuals may be positive or negative. The simple model we have constructed splits up the variability in our data into two parts: that which can be accounted for (expected value, which are our treatments), and that which cannot be accounted for (residual, which is the error). Residuals can be thought of as what is left (the residue) after we have done our best to account for the variation in our data. Another way of thinking of this is to consider that we can control some sources of variation (the treatments), but that leaves us with an uncontrolled source of variation (the residuals).

### 5.7.1 Blocking

There are four replicates of each treatment in our experiment, and the 16 plots were arranged at random with four of them on each side of the field to allow for the fact that the four sides of the field may well provide slightly different environments. We will now see how to take this information into account in a revised layout known as a randomised complete block design and so reduce random variation and improve our ability to detect differences between treatments (Figure 5.1 and Box 5.1).

## Box 5.1 Another Way of Presenting This Equation

Some textbooks use equations to represent the same idea:

$$y_{ij} = \bar{y} + t_i + e_{ij}$$

Letters $i$ and $j$ are called **subscripts**. The letter $i$ stands for the treatment number. In our experiment we could replace it by 1, 2, 3, or 4. The letter $j$ stands for the replicate number. We could use 1, 2, 3, or 4 here. So, if we wished to consider the number of spiders in treatment 2, replicate 3, we would replace these subscripts of $y$ (the observed value) by 2 and 3, respectively. The grand mean is represented by $y$ with a bar above it. The treatment effect is given by $t$ with a subscript $i$ that represents the appropriate treatment. In our case we would use 2 for treatment 2. Finally, the residual is represented by the letter $e$ (for error), which again has subscripts for the appropriate treatment and replicate because the random variation differs from plot to plot.

The letter $e$ is often used for the error or residuals, which indicates the unexplainable component of variance. Similarly, "fitted value" is often used in place of the expected value but means the same thing.

If we add a blocking factor to the experiment the extra term $b_j$ is simply added to the right hand side:

$$y_{ij} = \bar{y} + t_i + b_j + e_{ij}$$

In a randomised complete block design we can reduce the random variation in our data by accounting for variation between blocks as well as variation between treatments (Section 5.5.2). We now have an extra classification the block effect in our model:

*Observation = grand mean + treatment effect + block effect + residual*

Remember that the treatment effect is defined as the difference between treatment mean and grand mean. The block effect is defined similarly as the difference between the block mean and the grand mean.

Here are the observations of the number of spiders per plot, again. This time we have calculated both treatment and block means (Table 5.4).

**TABLE 5.4**

Block Means and Treatment Means

| Block | Treatment | | | | Mean |
|---|---|---|---|---|---|
| | F1 | F2 | NF1 | NF2 | |
| 1 | 21 | 16 | 18 | 14 | 17.25 |
| 2 | 20 | 16 | 17 | 13 | 16.5 |
| 3 | 19 | 14 | 15 | 13 | 15.25 |
| 4 | 18 | 14 | 16 | 12 | 15.0 |
| Mean | 19.5 | 15.0 | 16.5 | 13.0 | 16.0 |

The grand mean is shown at the bottom right, it is 16.0. This has not changed. We can now use our model to calculate a table of the same layout but showing expected values.

$$expected\ value = grand\ mean + treatment\ effect + block\ effect$$

The block and treatment means are as we calculated them from our data. We are now predicting the expected value for each plot from our model. Let us concentrate on treatment F1 in block 1 (top left in Table 5.4). To calculate the expected value we must know the grand mean (16) and both the treatment and block effects as before:

$$Treatment\_effect = treatmentmean - grandmean$$

$$= 19.5 - 16 = 3.5$$

Block effect is calculated in the same way as the treatment effect:

$$Blockeffect = blockmean - grandmean$$

$$= 17.25 - 16 = 1.25$$

$$Expectedvalue = grandmean + treatmenteffect + blockeffect$$

$$= 16 + 3.5 + 1.25 = 20.75$$

We can now start to construct a table of expected values.

Use the model to calculate the expected values for treatment F2 in Block 1 and for treatment F1 in block 2. You should obtain 16.25 and 20.0, respectively (Table 5.5a).

There is another way to calculate expected values once the first one has been calculated. It is quicker and sheds light on what the model is doing. On average, all plots in treatment F2 have 4.5 fewer spiders than those in treatment F1 (19.5 – 15.0). So, to get the expected value for treatment F2 in block 1, we take away 4.5 from the expected value for treatment F1 in block 1. This gives 16.25. Similarly, to get the expected value for treatment NF1 in

**TABLE 5.5a**

Calculation of Expected Value for Each Plot

| Block | Treatment | | | | Mean |
|-------|-------|-------|-------|-------|-------|
|       | **F1** | **F2** | **NF1** | **NF2** | |
| 1 | 20.75 | 16.25 | | | 17.25 |
| 2 | 20.0 | | | | 16.5 |
| 3 | | | | | 15.25 |
| 4 | | | | | 15.0 |
| Mean | 19.5 | 15.0 | 16.5 | 13.0 | 16.0 |

**TABLE 5.5b**

Expected Values per Plot

| Block | Treatment | | | | Mean |
|-------|-------|-------|-------|-------|-------|
|       | **F1** | **F2** | **NF1** | **NF2** | |
| 1 | 20.75 | 16.25 | 17.75 | 14.25 | 17.25 |
| 2 | 20.0 | 15.5 | 17.0 | 13.5 | 16.5 |
| 3 | 18.75 | 14.25 | 15.75 | 12.25 | 15.25 |
| 4 | 18.5 | 14.0 | 15.5 | 12.0 | 15.0 |
| Mean | 19.5 | 15.0 | 16.5 | 13.0 | 16.0 |

block 1, we add 1.5 (the difference between the treatment means for F2 and NF1) to the expected value for F2 in block 1 and get 17.75. The same idea works for calculating expected values in the same column. On average, all plots in block 1 have 0.75 more spiders than those in block 2. Therefore, the expected value for treatment F1 in block 2 is 20.75 – 0.75 = 20.0 and so on (Tables 5.5a and 5.5b).

Now we know what our model predicts. How good is it at explaining variation in spider numbers? If it was a perfect fit, we would find that the tables of observed and expected values (Table 5.2 and Table 5.5b) were the same as each other. This is very unlikely. Usually, there are differences, which are the residuals. We remember that

$$Residual = observed\ value - expected\ value\ \text{(Table 5.6).}$$

These residuals seem quite small. This suggests that our model is quite good at explaining variation in spider numbers between plots, and there is not much that cannot be explained either in terms of a treatment or a block effect. But how big do residuals have to be before you start being concerned that the model is not a good fit; how do you decide whether differences between the treatments are big enough to mean anything? After all, we carried out this experiment to answer the question "Is there any difference between the effects of the four treatments on spider numbers?"

In this chapter we have constructed a model to describe our experiment, which seems to be relevant. This is the first step towards answering our question, but in the next chapter we need to go a step further and decide

**TABLE 5.6**

Residuals per Plot

| Block | F1 | F2 | NF1 | NF2 | Mean |
|-------|-----|-------|-------|-------|------|
| | | Treatment | | | |
| 1 | 0.25 | -0.25 | 0.25 | -0.25 | 0 |
| 2 | 0 | 0.5 | 0 | -0.5 | 0 |
| 3 | 0.25 | -0.25 | -0.75 | 0.75 | 0 |
| 4 | -0.5 | 0 | 0.5 | 0 | 0 |
| Mean | 0 | 0 | 0 | 0 | 0 |

whether this model fits the data satisfactorily, and how to reach a preliminary conclusion on the significance of the apparent differences between treatments.

# 6

## Analysing Your Results: Is There Anything There?

> Figures often beguile me, particularly when I have the arranging of them myself.
>
> **Mark Twain**

In Chapter 5 we saw how to construct a model to fit our experimental design. Here we will use this model to analyse the data from our experiment.

## 6.1  Is Spider Abundance Affected by Treatment?

We want to know whether the variation in spider numbers between plots can be explained by the different treatment combinations applied to them. In other words, we are asking whether any differences between groups are statistically significant. The treatments were: F1 (wildflowers, cut once), F2 (wildflowers, cut twice), NF1 (no flowers, cut once), and NF2 (no flowers, cut twice).

The technique we will use to analyse our data and reach a conclusion is called *analysis of variance*, or ANOVA. It was published by R. A. Fisher in 1923, and it works by splitting up the variation in the data and determining how much of it is explained by each source of variation (treatment, block, and error) in our model. In Chapter 3 we introduced the t-test for comparing means of two different treatments. ANOVA uses the same approach but can also deal with more than two treatments.

## 6.2  Why Not Use Multiple t-Tests?

There are several reasons why we do not use t-tests to make every possible comparison. The most important is problem of **multiplicity of inference**. If we have four groups, then there are six different comparisons that could be made between any two groups. If we had conducted the tests using a rejection criterion of 5% as usual, the probability that each test gives us a Type I error, if the Null hypothesis is true, is 0.05, but the probability that at least one test gives a Type I error could be as high as

$$.05 + .05 + .05 + .05 + .05 + .05 = 0.05 \times 6 = 0.30$$

One way around this would be to conduct all the tests with a rejection criterion of

$$\frac{0.05}{6} = 0.0083333, \text{ or, in general, } \frac{\alpha}{N}$$

where $\alpha$ is the overall error rate we are willing to accept (usually 0.05), and N is the number of tests we can conduct. This is called the **Bonferroni method**, and it does have its uses, but here it would be a very conservative approach, mainly because it takes into account all six possible comparisons, several of which may not be of interest.

A second reason for avoiding multiple t-testing is the fact that we do not use all the information in each test. For example, if we compared treatment F1 with treatment F2 in the spiders example, we would ignore the information about the overall variation in plots assigned to the NF1 and NF2 treatments.

Finally, testing all possible pairs of means ignores the structure of the data. The best test of the effect of cutting regime would be a combined comparison of the difference between F1 and F2 *and* the difference between NF1 and NF2. Using ANOVA means that we can analyse all our data with one test, and if we design it correctly, we can make all the informative comparisons within the context of a single measure of overall variation.

---

## 6.3 ANOVA for a Wholly Randomised Design

For simplicity, we will start by assuming that our data (Table 6.1) came from a wholly randomised experiment with four replicates per treatment (Figure 6.1). Once we have shown how the approach works, we will include the idea of block variation. A wholly randomised design is often referred to as a "one-way" ANOVA because there is only one source of variation involved (error variation is assumed to be in every analysis).

We have three types of variation; the total, the treatment, and the error. We partition the total into what we can explain (the treatment) and what we cannot explain (the error), and compare them to see their relative importance. Figure 6.2 shows graphically the different types of variation. The standard

**TABLE 6.1**

The Observed Numbers of Spiders in Each Plot

| Replicate | Treatment | | | |
|---|---|---|---|---|
| | **F1** | **F2** | **NF1** | **NF2** |
| 1 | 21 | 16 | 18 | 14 |
| 2 | 20 | 16 | 17 | 13 |
| 3 | 19 | 14 | 15 | 13 |
| 4 | 18 | 14 | 16 | 12 |
| Mean | 19.5 | 15.0 | 16.5 | 13.0 |

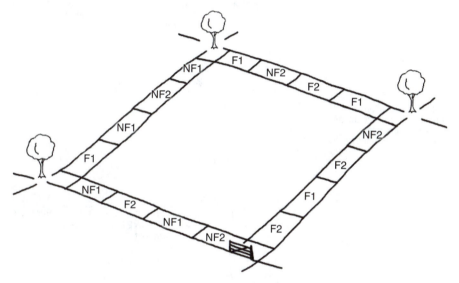

**FIGURE 6.1**
The field experiment, with the treatments randomly allocated across all plots rather than into blocks.

way to express each of these sources of variation is to calculate the square of each deviation shown in the figures and to add them all up to get "sums of squared deviations" or "sums-of-squares" for short.

## 6.3.1 Calculating the Total Sum-of-Squares

R. A. Fisher's great discovery was that the sums of squared deviations add together in a way that reflects the different sources of variation:

*Total sum-of-squares = treatment sum-of-squares + residual sum-of-squares*

First, we calculate the total sum-of-squares, the variation for all the data taken together. It is the sum of the squares of the differences between each observation and the grand mean. We discussed this as a measure of variation

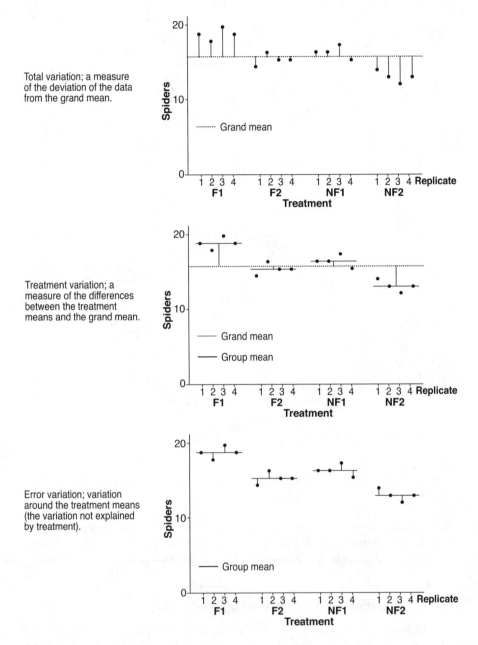

**FIGURE 6.2**
A diagrammatic representation of the different components of variation.

in Chapter 1, Section 1.6.2. The data for spiders is given, again, in Table 6.1. Box 6.1 gives a manual method for doing the calculations. We get 106.0. We can also do this in MINITAB even more simply, and we will cover this at the end of this chapter.

---

## BOX 6.1 *Don't Panic!* How Computers Calculate Sums-of-Squares

It is our general contention in this book that calculation is a job for computers but that working through the calculations manually can be helpful at the start. The direct way to get a sum of squared deviations is to work out the deviations, square each one, and add them all up. So, for the total sum of squares we would have to calculate:

$$(21 - 16)^2 + (20 - 16)^2 + \ldots$$

for all 16 plots. Quite apart from being extremely tedious, it is quite likely we would make a mistake!

A better method relies on the fact that most calculators have a "standard deviation" program. In Chapter 1 we saw that, in mathematical terms, the sample standard deviation is defined as

$$s = \sqrt{\frac{(x - \bar{x})^2}{(n-1)}} \, ,$$

where the term $(x - \bar{x})^2$ is precisely the sum of squared deviations we are after; let us write it as $SSX$ to make things simpler. We then use a simple bit of algebra to show that

$$s = \sqrt{\frac{SSX}{(n-1)}}$$

$$s^2 = \frac{SSX}{(n-1)}$$

$$s^2 \times (n-1) = SSX$$

To use this on the calculator, all you need to do is

- Enter all 16 observations (Table 6.1) into our calculator on statistical mode.
- Press the standard deviation button $\sigma(n-1)$.
- Square it to get the variance.
- Multiply this by $n-1$ (15) to get the total sum-of-squares.

The answer should be 106.0.

### 6.3.2  Calculating the Error (Residual) Sum-of-Squares

We have just calculated the total sum-of-squares; therefore, if we can calculate the **error** (or residual) sum-of-squares, we can find out the treatment sum-of-squares by subtraction. We want this because it tells us how much of the total variation is due to treatments. First, we need to know what the "residuals" are for each observation. As we saw in the last chapter, they are the differences between the observed and expected values. The expected values are the group means, so using the observed and expected values, we can calculate the residuals. Figure 6.3 shows how to do this for treatment F1. We would then do the same for the other three treatments and add the four sums of squared residuals together.

### 6.3.3  Calculating the Treatment Sum-of-Squares

*By subtraction:* There is only one source of controlled variation in this experiment (each of the four treatments we have allocated), so using the equation below, we can calculate the treatment sum-of-squares as the difference between the total sum-of-squares and the error sum-of-squares.

$$Total\ sum\text{-}of\text{-}squares = treatment\ sum\text{-}of\text{-}squares + error\ sum\text{-}of\text{-}squares$$
$$106.0 = treatment\ sum\text{-}of\text{-}squares + 16.0$$

Therefore,

$$Treatment\ sum\text{-}of\text{-}squares = total\ sum\text{-}of\text{-}squares - error\ sum\text{-}of\text{-}squares$$
$$= 106.0 - 16.0 = 90.0$$

Each residual is the observed − expected.

| Replicate | Observed | Expected | Residuals | Squared Residuals |
|-----------|----------|----------|-----------|-------------------|
| 1 | 21 | 19.5 | 1.5 | 2.25 |
| 2 | 20 | 19.5 | 0.5 | 0.25 |
| 3 | 19 | 19.5 | -0.5 | 0.25 |
| 4 | 18 | 19.5 | -1.5 | 2.25 |
| sum | 78 | 78 | 0 | 5 |

When the residuals are squared, they do not sum to zero.

The expected value is the mean for the treatment, so all values are the same in this column.

The residuals sum to zero.

**FIGURE 6.3**
Observed values, expected values, and residuals for treatment F1.

*From first principles:* The preceding formula only works when there is one source of controlled variation in the experiment. If there is more than one, such as when we include blocking in our design, we need to calculate the treatment sum-of-squares for each of these sources, as they will explain different amounts of variation.

The first stage is to look at how the treatment means vary. Our four treatment means are 19.5, 15.0, 16.5, and 13.0. To get the treatment sum-of-squares we enter these figures into the calculator on statistical mode. We then calculate their variance (using the short cut noted earlier) by pressing the standard deviation button $(n - 1)$ and squaring it. We then multiply by $n - 1$ (3). This gives us the sum-of-squares of these four numbers (the treatment means), but it ignores the fact that each group contains four replicate observations for which *"group mean–grand mean"* is the treatment effect. To obtain the treatment sum-of-squares on a "per plot" scale (similar to the total and residual sums-of-squares), we must multiply our result by 4 (the number of observations in each treatment total).

You should find that the answer is 90. This is in agreement with the result we obtained by subtracting the residual sum-of-squares from the total sum-of-squares.

## 6.4 Comparing the Sources of Variation

When we compare the sources of variation, we find that, in this experiment, the treatments account for most of the total variation. The experiment has been reasonably successful in that other sources of variation have been kept to a minimum. This will not always be the case, so let us consider the two extreme situations we could encounter where treatments explain nothing or everything (Figure 6.4).

## 6.5 The Two Extremes of Explanation: All or Nothing

### 6.5.1 Treatments Explain Nothing

If all the treatment means are the same, knowing the relevant treatment mean for a plot does not help you to predict anything specific about it. In this case, the error variation equals the total variation, and the treatment variation would be zero (Figure 6.4a).

Here the total sum-of-squares equals the sum-of-squares of the residuals. The treatment sum-of-squares is zero, so it looks as if we have just randomly picked numbers and assigned them to treatments.

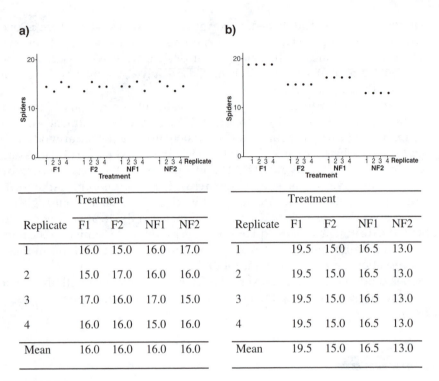

| | Treatment | | | | | Treatment | | | |
|---|---|---|---|---|---|---|---|---|---|
| Replicate | F1 | F2 | NF1 | NF2 | Replicate | F1 | F2 | NF1 | NF2 |
| 1 | 16.0 | 15.0 | 16.0 | 17.0 | 1 | 19.5 | 15.0 | 16.5 | 13.0 |
| 2 | 15.0 | 17.0 | 16.0 | 16.0 | 2 | 19.5 | 15.0 | 16.5 | 13.0 |
| 3 | 17.0 | 16.0 | 17.0 | 15.0 | 3 | 19.5 | 15.0 | 16.5 | 13.0 |
| 4 | 16.0 | 16.0 | 15.0 | 16.0 | 4 | 19.5 | 15.0 | 16.5 | 13.0 |
| Mean | 16.0 | 16.0 | 16.0 | 16.0 | Mean | 19.5 | 15.0 | 16.5 | 13.0 |

**FIGURE 6.4**

Two extremes of results. In (a) the treatments explain nothing, so the group means are the same, and there is only random variation. The opposite is true of (b) where treatments explain everything. Here the group means are very different, and there is no error variation.

### 6.5.2   Treatments Explain Everything

At the other extreme, if treatment effects explained everything, then the treatment variation would account for all the total variation, and the random variation (the error variation) would be zero. In this situation, if you know the treatment mean, you can predict the plot mean perfectly (Figure 6.4b). There is no random variation within any treatment. All residuals are zero because the observed and expected values are equal in each plot. The total sum-of-squares equals the treatment sum-of-squares. The sum-of-squares of the residuals (also called the residual sum-of-squares) is zero.

## 6.6   The ANOVA Table

### 6.6.1   Sources of Variation and Degrees of Freedom

Whatever way we have calculated the treatment sum-of-squares. We now adopt a standard method of presentation as follows. We start by writing

**TABLE 6.2**

The Sources of Variation and Their Degrees
for Freedom for the Field Experiment

| Source of variation | d.f. |
|---|---|
| Treatments | 3 |
| Error | 12 |
| Total | 15 |

down the analysis of variance plan. This gives the structure of the experiment in terms of the sources of variation, together with their degrees of freedom (Table 6.2).

From now on, we will refer to the "residual" as "Error," as this is what most statistical computer packages call it. The total variation is always on the bottom line with the Error immediately above it. Sources of controlled variation (Treatments) appear above them. There are 15 degrees of freedom (d.f.) for the Total line because there are 16 plots, and d.f. = n – 1. Similarly, there are 3 d.f. for treatments because there are four treatments, so n – 1 = 3. Error d.f. are obtained by subtracting 3 d.f. from 15 d.f. Another way of thinking about error d.f. is to spot that each of the four treatments has four replicates. So, within each treatment, there are n – 1 = 3 d.f., which represent random variation. If error variation has 3 d.f. for each of the four treatments, this gives 12 d.f. in all.

It is very important to write down an analysis of variance plan before carrying out any experiment. Otherwise, we may find out too late that the experiment either cannot be analysed, has too little replication to be able to detect differences between population means, or has excessive replication and so is inefficient. If your experiment is complicated, the simplest way to do this is to simulate random data by inputting made-up numbers into a spreadsheet with the appropriate treatments. Try to analyse it; if the computer cannot produce an ANOVA table, something is wrong with your design and you need to think again (cf. Section 4.5).

### 6.6.2 Comparing the Sums of Squares

We can now fill in the next part of the analysis of variance table, Figure 6.5. First, we put in the sums-of-squares calculated above (Section 6.3). We again see the additivity — the total sum-of-squares is the sum of all the numbers above it in the sum-of-squares column. In this case, the sum-of-squares for treatment seems much larger than that for error, suggesting that treatments explain a substantial part of the variation in spider numbers; in fact, to be precise, it is $90/106 = 0.849$ or 84.9% of the variation. This quantity is a useful measure of how good the model is, and it is called (for mysterious reasons we will reveal in Chapter 8) "R-squared" or $r^2$.

| | | Amount of variation explained | Variance (sum-of -squares per degree of freedom) | Ratio comparing the variance we can explain (treatment) to the variance we cannot explain (error) |
|---|---|---|---|---|
| Number of independent pieces of information | ↓ | ↓ | ↓ | ↓ |
| Source | df | Sum-of-squares | Mean square | F=ratio |
| Treatments | 3 | 90.0 | 30.0 | 22.5 |
| Error | 12 | 16.0 | 1.3333 | |
| Total | 15 | 106.0 | | |

F-ratio = Treatments mean square / Error mean square

**FIGURE 6.5**
The components of the ANOVA table and how the F-ratio is calculated.

### 6.6.3  The Mean Square

There is a problem with using the sum-of-squares comparison to decide about hypotheses. Its magnitude depends on the number of treatments and the number of replicates. We control for this by dividing the sums-of-squares by the associated degrees of freedom to give a measure of variation per degree of freedom. This new measure, the **mean square,** is comparable between analyses and experiments. In fact, it is a **variance**; if you look back at Chapter 1 or at Box 6.1, you can see that a variance actually is a sum of squared deviations about a mean divided by the relevant degrees of freedom.

We only need to work out the mean square for the treatment and residual lines (as these are what we will compare). So, the treatment mean square = 90.0/3 = 30.0, and the error mean square = 16.0/12 =1.33. These are filled in the next column.

The treatment and error mean squares can now be compared directly because both are expressed in terms of the number of independent pieces of information they contain (the degrees of freedom). If the treatment mean square were zero, we would be in the situation described in Section 6.5.1: "treatments explain nothing." If the residual mean square were zero, treatments would explain everything as in Section 6.5.2. We can quantify where we are between these two extremes by dividing the treatment mean square by the residual mean square. This gives a ratio of two mean squares (or variances). It is called the **variance ratio** or **F-ratio,** and is the *test statistic* we need for hypothesis tests about the effect of the treatment. If treatments explained nothing, the F-ratio will be 0. If our observations had been selected from one population and allocated to treatments at random, we would expect an F-ratio of 1 as treatment and error mean squares would be equal. If treatments explained everything, the F-ratio will be infinity (very large!). The larger our variance ratio, the more evidence we have that the treatments differ from each other more than just by chance. Here we get a value of 22.5 (30.0/1.3333).

## 6.7   Testing Our Hypothesis

As we indicated in Chapter 5, Section 5.3, the model *spiders = treatment* corresponds to the research hypothesis that at least one of the treatments leads to different numbers of spiders from the others. The Null hypothesis under test is, therefore, that all groups have the same mean number of spiders.

The reasoning behind the test is the same as we saw in Chapter 2. We calculate how often we would get this spread of results if treatments have no effect whatsoever.

In practice, all we need to do is see how to use our calculated test statistic (F) to establish whether or not it is safe to reject the Null hypothesis. As with the t-test, the distribution of F is known in the case that the Null hypothesis is true and can be found in Table C.3 to Table C.5 in Appendix C in the back of the book.

Different rejection criteria are represented in separate tables: one for 5%, one for 1%, and one for 0.1% rejection probability. Recall from Chapter 2 that these probabilities (known as α) represent the probability of a Type I error (rejecting the Null hypothesis when it is true) given *that it is in fact true.* Thus, the first step in deciding our rejection criterion is to choose the value of α we wish to work with. Conventionally, it is usually the 5% level, so we use Table C.3.

Like the t-distribution, degrees of freedom are important, but instead of one set of d.f. there are now two. This is because F is the **ratio** of two mean squares, each of which has degrees of freedom. It is important to keep track of which are which. Conventionally, the d.f. in the column headings are labelled $n_1$ and those for the rows are labelled $n_2$. An F-ratio is considered to be calculated from the mean square for the effect of interest (Treatment, in our example) with $n_1$ d.f. divided by the error mean square with $n_2$ d.f. The body of Table C.3 contains the F-ratios such that they would be exceeded only 5% of the time *if the Null hypothesis in question is true.* In our example, the treatment degrees of freedom are 3 (n groups – 1) and the error degrees of freedom are 12, so the critical value for F is 3.49 (Figure 6.6).

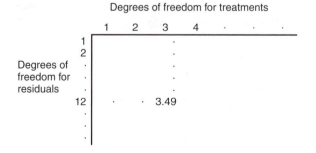

**FIGURE 6.6**
Looking up the critical F value with the degrees of freedom.

Our F-ratio of 22.5 is much greater than this. Therefore, we may conclude that we have evidence to reject the Null hypothesis because we would get an F-ratio bigger than 3.49 less than 5% of the time if the Null hypothesis were true. We could represent this as $p < 0.05$ in an account of the analysis, but this F-ratio is so much bigger than 3.49 that our p-value is likely to be much smaller. We can look this up in Table C.4 or Table C.5. The critical values for $\alpha = 1\%$ with 3,12 d.f. is 5.95 (Table C.4), and for $\alpha = 0.1\%$ it is 10.8 (Table C.5). The calculated value exceeds the 0.1% critical value, so we can say that the result is significant with $p < 0.001$. We therefore conclude that there is very strong evidence to reject the Null hypothesis. The probability $p < 0.001$ tells us that there is less than a one-in-a-thousand chance of us obtaining our particular set of experimental results if the treatments really all have the same effect (and so, the observations really come from only one population). This is such a small chance that we would conclude that the treatments do not all have the same effect on spiders.

We write this as $F_{3,12} = 22.5; P < 0.001$.

Note that we include the degrees of freedom for the F-ratio so that other scientists have an idea of our replication. One advantage of using a computer is that we can get an exact value for the probability rather than just comparing the calculated F with a critical value.

## 6.8  Including Blocks: Randomised Complete Block Designs

When we described the spiders experiment in Chapter 5, we said that we should be using a blocking structure to deal with the fact that field margins on different sides of the field might have different numbers of spiders irrespective of the treatment. We will now see how to include this in the analysis. We introduce an extra line to the ANOVA table to represent the amount of variation accounted for by the blocks. This is often called a *two-way ANOVA* because there are two classifications: treatments and blocks. By accounting for variation caused by differences within the site through blocking, we are likely to reduce the random or error variation. In this case, the error mean square will decrease and the variance ratio for treatment effects will increase.

First, let us consider the analysis of variance plan including blocking (Table 6.3).

**TABLE 6.3**

Sources of Variation When Block Is Included and Their Degrees of Freedom

| Source of Variation | d.f. |
| --- | --- |
| Blocks | 3 |
| Treatments | 3 |
| Error | 9 |
| Total | 15 |

Because there are four blocks, there are 3 d.f. for blocks. We add together 3 d.f. for blocks and 3 d.f. for treatments and subtract the answer (6) from the total d.f. (15) to obtain the d.f. for the error (9). This is 3 fewer than it was before because 3 d.f. are now accounted for by blocks. Just as the old d.f. for the error has now been split into the d.f. for blocks and the new d.f. for error, so the sum-of-squares for the error will be split; part will now belong to blocks and the remainder will be the new, smaller, error sum-of-squares.

We must now calculate a sum-of-squares for blocks. The method is just the same as for treatments. Our four block means are 17.25, 16.5, 15.25, and 15.0. Using the method in Box 6.1 to get the block sum-of-squares, we enter the block means into the calculator on statistical mode. We press the standard deviation button (n – 1) and square it to get the variance. We then multiply by n – 1 (3). This gives us the sum-of-squares of these four means but, as with the treatment sum-of-squares, it ignores the fact that each of them is itself derived from four observations (one plot from each treatment). To obtain the block sum-of-squares on a per-plot scale (similar to the total, treatment, and error sums-of-squares), we must multiply our result by 4 (the number of observations in each block total).

Here is the working:

$$\text{Block totals } 17.25, 16.5, 15.25, 15.0$$
$$\text{standard deviation (of block means)} = 1.061$$
$$\text{variance} = 1.125$$
$$\text{variance} * 3 = 3.375$$
$$3.375 * 4 = 13.5 = \text{block sum-of-squares}$$

We now include the blocks sum-of-squares in the ANOVA table (Figure 6.7), and obtain the revised error sum-of-squares by subtracting 13.5 and 90.0 (treatment sum-of-squares) from 106.0 to give 2.5. Notice again that the treatment sum-of-squares and the total sum-of-squares remain the same. What has happened is that the error sum-of-squares from the previous analysis has split into two components: that due to blocks (13.5) and that due to other sources of variation not accounted for (2.5).

The F-ratio is calculated as the Mean square for the term of interest divided by the Error mean square.

| Source | df | Sum-of-squares | Mean square | F = ratio |
|---|---|---|---|---|
| Blocks | 3 | 13.5 | 4.5 | 16.2 |
| Treatments | 3 | 90.0 | 30.0 | 108.0 |
| Error | 9 | 2.5 | 0.278 | |
| Total | 15 | 106.0 | | |

$$16.2 = \frac{4.5}{0.278}$$

$$108.0 = \frac{30.0}{0.278}$$

**FIGURE 6.7**
Calculating the F-ratio with more than one term in the model.

The blocks mean square is obtained by dividing its sum-of-squares by degrees of freedom as before. The revised error mean square is then obtained by dividing 2.5 by 9. The revised variance ratio for treatments is obtained by dividing the treatments mean square by the revised error mean square to give 108.0. This value is then compared with the critical value in F tables for $n_1 = 3$ and $n_2 = 9$ d.f. Our variance ratio is very much bigger than the critical F value for $p = 0.001$. Therefore, we have strong evidence for rejecting the Null hypothesis.

The block mean square (4.5), though smaller than that for treatments, is also very much bigger than the error mean square. Therefore, we have strong evidence that the blocks are not merely random groups. They have accounted for site variation well, in that plots in some blocks tend to have more spiders in them than plots in other blocks. This source of variation has been identified and separated from the residual. Thus, the amount of random variation that remains unaccounted for is quite small. Figure 6.8 summarises the steps we have gone through to create an ANOVA table identifying two sources of variation.

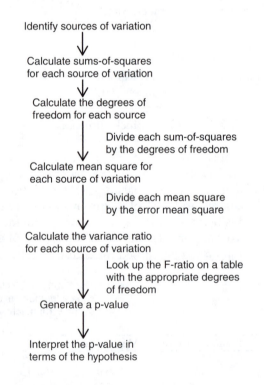

**FIGURE 6.8**
Flow chart to show how ANOVA goes from data to p-value.

## 6.9 Analysing the Spider Data Set in MINITAB

The experiment studied the effect of four different combinations of sowing and cutting on the number of spiders. We type our data and block number and treatment code (see the following text) into columns in the MINITAB worksheet, check and edit it, and then save it onto a disc. We can then ask MINITAB to print out a copy of our information:

```
MTB > print C1-C3
```
Note: MINITAB provides the row number on the left-hand side (column 0).

| Row | Treatment | Block | Spiders |
|-----|-----------|-------|---------|
| 1 | F1 | 1 | 21 |
| 2 | F1 | 2 | 20 |
| 3 | F1 | 3 | 19 |
| 4 | F1 | 4 | 18 |
| 5 | F2 | 1 | 16 |
| 6 | F2 | 2 | 16 |
| 7 | F2 | 3 | 14 |
| 8 | F2 | 4 | 14 |
| 9 | NF1 | 1 | 18 |
| 10 | NF1 | 2 | 17 |
| 11 | NF1 | 3 | 15 |
| 12 | NF1 | 4 | 16 |
| 13 | NF2 | 1 | 14 |
| 14 | NF2 | 2 | 13 |
| 15 | NF2 | 3 | 13 |
| 16 | NF2 | 4 | 12 |

### 6.9.1 One-Way ANOVA

This assumes that the treatments were allocated at random and not blocked, and so ignores the block codes in column 2. The menu commands below, ask MINITAB to perform an ANOVA using the model Spiders = Treatment, and to print the means of the treatments. The output is shown below.

**Stat>ANOVA> Balanced ANOVA.**
**Responses:** Spiders **Model:** Treatment.
**Results> Display means corresponding to the terms:** Treatment.

**ANOVA: Spiders vs. Treatment**

| Factor | Type | Levels | Values |
|--------|------|--------|--------|
| Treatment | fixed | 4 | F1, F2, NF1, NF2 |

```
Analysis of Variance for Spiders

Source       DF       SS       MS       F       P
Treatment     3   90.000   30.000   22.50   0.000
Error        12   16.000    1.333
Total        15  106.000

Means

Treatment   N  Spiders
        1   4   19.500
        2   4   15.000
        3   4   16.500
        4   4   13.000
```

MINITAB responds by first summarising the model: the factor "Treatment" has four levels, which are F1, F2, NF1, and NF2. MINITAB produces an ANOVA table in the conventional format (as Figure 6.7). The p-value (0.000) is extremely small ($< 0.001$). So, the chance of getting our sample of results if all four treatments are samples from the same population is extremely unlikely (less than one chance in 1000). We conclude that the treatments do not all have the same effect on spider numbers.

### 6.9.2   Two-Way ANOVA

Now we introduce blocking into the model. Thus, there are now two sources of variation that we can identify: treatments and blocks.

**Stat>ANOVA>Balanced ANOVA**
**Model:** Treatment Block
**Results>Display means corresponding to the terms:** Treatment Block

**ANOVA: Spiders vs. Treatment, Block**

```
Factor       Type   Levels  Values
Treatment    fixed       4  F1, F2, NF1, NF2
Block        fixed       4  1, 2, 3, 4

ANOVA for Spiders

Source       DF       SS       MS       F       P
Treatment     3   90.000   30.000   108.00   0.000
Block         3   13.500    4.500    16.20   0.001
Error         9    2.500    0.278
Total        15  106.000
```

```
Means

Treatment   N   Spiders
         1  4    19.500
         2  4    15.000
         3  4    16.500
         4  4    13.000

Block   N   Spiders
    1   4    17.250
    2   4    16.500
    3   4    15.250
    4   4    15.000
```

The variance ratio for treatments (F = 108) is now much higher than it was in the one-way analysis. We have removed variation between blocks from the random variation (error). The treatment mean square is therefore now compared with a much smaller mean square for randomness (0.278) giving us even stronger evidence for rejecting the Null hypothesis of no difference between the effects of the four treatments.

### 6.9.3  Box Plots for Treatments

We can ask MINITAB to display the results for each treatment as a box plot (Figure 6.9). We can see how plots receiving treatment F1 have a high number of spiders, whereas those receiving treatment NF2 have a low number.

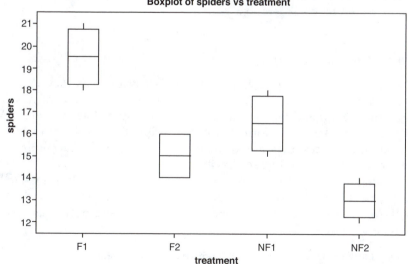

**FIGURE 6.9**
Box plot to show the mean and interquartile range of each treatment.

## 6.10 The Assumptions Behind ANOVA and How to Test Them

When we carry out an ANOVA, we have to make certain assumptions about the nature of the observations and about the underlying distributions. If these assumptions are not met, then the relationship between the F-statistic values and the probability they will occur if the Null hypothesis is true breaks down, so that the risk of a Type I error when we reject $H_0$ may be much larger than we thought it was. The assumptions are: independence, Normality of error, homogeneity of variance, and additivity.

### 6.10.1 Independence

It is vital that the replicates of each treatment were *independent* of each other and that the experiment was *randomised*. If some of the measurements depend on one another (e.g., if we counted each quadrat twice "just to be sure" and included both measures in the analysis), then the relationship between the variation in the measurements and their number is distorted. In that case, the overall (error) variation would be smaller than it should be for the number of observations we claimed to have made. This would mean that the error mean square would be *too small*, and hence the F-ratio *too big* for the error degrees of freedom we are claiming. As a result, the risk of a Type I error would be greater than we thought.

There is no way to test whether the independence assumption has been met other than by thinking carefully about the design of the study and ensuring that each measurement is a "fair and independent test" of the hypotheses under consideration.

### 6.10.2 Normality of Error

A Normal distribution of residuals (errors) is assumed, as they represent random variation. There should be many residuals with very small absolute values (near zero) and only a few with very large ones (far from zero). We test this by checking the Normality of the residuals from the model. Note that, provided the residuals are Normally distributed, there is no requirement that the raw data should be so distributed.

Checking the distribution of residuals when doing the analysis by hand is scarcely practical, but most statistical packages offer a quick way to do it. In MINITAB we can ask for a histogram of residuals and a Normal plot of residuals as shown:

```
Stat>ANOVA>Balanced ANOVA Responses: Spiders Model:
Block+Treatment
Storage: fits residuals.
Graphs: Histogram of residuals, Normal plot of residuals.
```

In Figure 6.10 we can see that the residuals look approximately Normal, and the Normality plot (Figure 6.11) approximates a straight line. These "eyeball" tests are usually considered sufficiently accurate, but if a formal test is felt to be necessary, the residuals must be saved in the worksheet, as shown under the Storage option, and then subjected to a formal test for Normality (we will not cover this here).

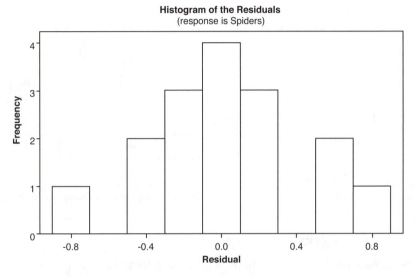

**FIGURE 6.10**
Histogram of residuals used to check the assumption that the residuals are Normally distributed.

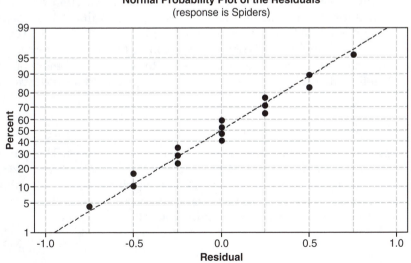

**FIGURE 6.11**
Normality plot; another way to check the assumption of Normality of error.

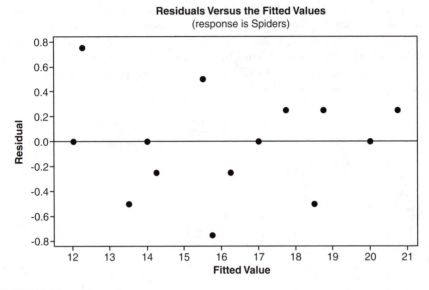

**FIGURE 6.12**
Residuals vs. fits, which is a diagnostic tool to check the homogeneity of variance assumption.

### 6.10.3    Homogeneity of Variance

We also assume that the variability within each treatment is similar. This is because we use its average (the residual mean square) to represent random variation. If this is much too small for some treatments and much too large for others, we may find that the F-ratio is larger than it should be with the consequent risk of a Type I error. We can examine this assumption by plotting the residual for each plot against its fitted or expected value. We just check the "Residuals vs. fits" option under the "Graphs" button (Figure 6.12).

The residuals should be distributed approximately evenly across the plot on either side of the x-axis with no tendency to bunch or fall more on one side than the other. The example in Figure 6.12 appears to be satisfactory.

### 6.10.4    Additivity

It is assumed that in our final model the effects are additive. Thus, in the blocked analysis it should be true that the effects of each treatment are the same relative to the others in each block. We will have more to say about this assumption in the next chapter; we have not yet covered the methods we need to test the additivity assumption formally.

### 6.11 Another Use for the F-Test: Testing Homogeneity of Variance

In the specific case of just two groups, which is in fact the same as an independent samples t-test, we can use the F-tables to formally check whether the variance is homogeneous. The procedure is to calculate the variance ($sd^2$) of each of the two groups and divide the larger by the smaller. The result is an F-ratio with $n_1 - 1$, $n_2 - 1$ degrees of freedom where $n_1$ is the sample size of the group with the larger variance, and $n_2$ that of the group with the smaller.

In our sheep diet example (Section 3.5), we found that the s.d. for the new diet group was 0.828 whereas that for the old diet group was 0.565. Remembering that the variance is just the standard deviation squared, we calculate:

$$F = \frac{0.828^2}{0.565^2} = 2.15 \ \ \text{(to 3 significant figures)}$$

The 5% critical value for F with $n_1 - 1$, $n_2 - 1 = 6, 6$ d.f. is 4.28, so we do not reject the Null hypothesis that the variances are equal, and we can confidently carry out a t-test (or ANOVA if we prefer it) on these data. Note that the rule of thumb we gave in Chapter 3, Section 3.8.2 — that the ratio of variances should be less than about 2.5 — is bit conservative for small samples, so it is probably worth using a formal test if the ratio exceeds 2.5 but sample sizes are less than about 30.

# 7

# *Interpreting Your Analysis: From Hypothesis Testing To Biological Meaning*

Sir Humphrey: "I strongly advise you not to ask a direct question."
Jim Hacker: "Why?"
Sir Humphrey: "It might provoke a direct answer."
Jim Hacker: "It never has yet!"

**Yes, Minister**
*A popular 1970s U.K. TV series*

The hypothesis testing method described in Chapter 6 is something of a blunt instrument. The Null hypothesis under test is rather broad: "Are the means of all the treatment levels the same as each other?" When it is clear that this global hypothesis should be rejected, how do we make sense of this result in terms of the research question "what effects do the treatments have"? In this chapter we will look at how we interpret differences between treatments. Then we will move towards using factorial experiments to tease out the biological effects in a more meaningful way. We will also expand on what we can do with ANOVA to investigate more complicated relationships.

We will do this by looking at estimates of treatment means, placing confidence intervals around them, and then looking at how to investigate interactions between two factors in our analysis. Finally we will see how a careful choice of the treatments allows us to obtain more specific and unambiguous information about the nature of the effects.

## 7.1   Treatment Means and Confidence Intervals

We continue with the field margins experiment of Chapter 6. We have established that the treatment means are significantly different. To look at the differences between them, we can plot the values to see how they compare, but we also need to know how confident we can be in these parameter estimates.

As we saw in Chapter 2, a confidence interval is the range within which we think the true value of an estimated parameter lies. We limit this to 95% of the time because the whole range could be enormous; 95% confidence also gives us a good compromise between Type I errors (finding that a Null hypothesis value is outside the confidence interval when it should not be) and Type II errors (finding a Null hypothesis value is inside the confidence interval when it should not be). Review Chapter 2 if you have forgotten about this.

We can calculate the upper and lower limit of the 95% confidence using the formula

$$confidence\ interval = estimated\ mean \pm (t_{crit} \times standard\ error\ of\ the\ mean)$$

The term in brackets contains the standard error, so that our confidence interval is linked to the spread of the results. This is multiplied by $t_{crit}$, which is a critical value for the error degrees of freedom. In this respect it resembles a t-test. (Chapter 2, Section 2.2).

The analogy persists when we think about what estimate to use for the population standard deviation . It may seem odd to speak of "the population standard deviation" when, in fact, we are comparing samples from four different populations. However, what we are proposing is that *if the Null hypothesis is true*, all four samples were drawn from the *same* population. Thus, we have four different estimates of the population standard deviation $\sigma$, namely the sample standard deviations of the four samples. Which should we use?

This is precisely the problem we met when considering the two-sample t-test in Chapter 3. The answer is "Use all of them." In the case of the t-test, this means forming the "pooled estimate" of the population standard deviation using Equation 3.1 in Box 3.2. In ANOVA, the square root of the error mean square is, in fact, just such a pooled estimate of the population standard deviation *provided the samples are drawn from populations with approximately similar variances*. (In fact, if we had just one factor with only two levels, the error mean square would be precisely the variance of the difference between the group means and is thus the square of the pooled standard deviation calculated in Box 3.2, Equation 3.1.) The "assumption of similar variances" for the t-test is just a special case of the assumption of "homogeneity of variance" in Chapter 6, Section 6.10.3. We check variances are the same between groups by plotting the data and seeing if the spread of results is the same between groups.

### 7.1.1   Calculating the 95% Confidence Interval for the Treatment Mean

We will calculate the 95% confidence interval for treatment F1 in the field margin experiment. The estimated value of the mean of F1 is 19.5 spiders per plot.

The value of t is looked up in t-tables for the error degrees of freedom. There are 12 d.f. for the error in the ANOVA table. We are interested in the 95% confidence interval (p = 0.05), so we look up the value under 12 d.f.

(for an ANOVA without blocking, see Chapter 6, Section 6.9 and Table 6.3) on the 95% t-table. It is 2.18.

Now, we need the standard error of the mean. As proposed above, we use the error mean square (= variance) from the whole experiment. This represents the pooled or average random variation within all treatments, after accounting for differences between them. If we divide the error mean square by the number of replicates of a particular treatment, we get the variance of the mean of that treatment. Then we take the square root to give the standard error of the mean. Here is the working:

Treatment mean = 19.5
t = 2.18 (for 12 degrees of freedom, in Table C.2).
Error mean square = 1.333 = variance (Figure 6.5)
n = number of replicates of the treatment = 4
Variance of the treatment mean = variance/n = 1.333/4 = 0.333
Standard error of the mean = square root of 0.333 = 0.577

Therefore, the confidence interval is

$$19.5 \pm (2.18 \times 0.577) = 19.5 \pm 1.26, \text{ namely, from 18.24 to 20.76}$$

There is a 95% chance that the true mean for treatment F1 lies between 18.3 and 20.8, and thus there is only a 5% chance that the mean is larger than 20.8 or smaller than 18.3, based on the evidence from this experiment.

## 7.2 Difference Between Two Treatment Means

As mentioned in Chapter 6, Section 6.2, in a complicated experiment, we want to carry out one test rather than many. This way we avoid the risk of some tests being significant purely because we have carried out too many of them. There is no point in carrying out an ANOVA and then testing all the pairwise comparisons afterwards. However, we could have a secondary aim of testing a hypothesis that there should be a difference between two specific treatments. This is alright, provided we decided this in advance, and we do not carry out several speculative tests.

We might want to calculate the confidence interval of a specific difference. For example, we are interested in the difference between treatment NF2 (mean = 13) and treatment F1 (mean = 19.5). In this case, the formula is

*confidence interval = difference ± (t × standard error of the difference)*

The difference is 6.5. Look up t with the error d.f. (12). This is 2.18, as before.

The standard error of the difference has to take into account the variance of the first mean and the variance of the second mean. This is because if we are estimating how far apart two population means are, and we have uncertain estimates of each of them, both amounts of uncertainty will have an effect on the range of size of the difference.

Both treatments have four replicates, so in this case the two treatment variances are the same as each other. We sum two variances of the means together:

$$0.333 + 0.333 = 0.666$$

We then take the square root to give 0.816 as the standard error of the difference between the two means.

We put the values into the formula for a 95% confidence interval

$$6.5 \pm (2.18 \times 0.816) = 6.5 \pm 1.78, \text{ or from } 4.7 \text{ to } 8.3$$

So, there is a 95% chance that the mean difference in spider numbers between the two populations lies between 5.7 and 7.3, based on the evidence from these data. This can be put another way: there is no more than a 5% chance that the mean difference lies outside this range, given these data. Thus, with these results, it is very unlikely that the mean difference is really zero, and we can therefore be very confident that the difference between these two treatments is real and not caused by chance. It is important to remember that we should not test all pairwise comparisons because of the problems of running too many tests discussed in Chapter 6, Section 6.1. We should only carry out a test like this if we have identified a specific comparison of interest when we designed the experiment.

## 7.3    Getting More Out of an Experiment: Factorial Designs and Interactions

We have already hinted why it might be more efficient to investigate several hypotheses in the same experiment if we use ANOVA. These types of experiments are called **factorial designs** and are very useful because of **hidden replication** and the ability to investigate **interactions** between **factors** (things we are investigating). Factorial designs save time and money by carrying out one experiment in which two factors are explored instead of examining each separately. Above all, we have the opportunity to find out whether these factors act independent of each other or whether they interact. If they do not interact, life is simple, and we can make general recommendations such as "We should sow flower seed to increase spider numbers." If they do interact, however, it is important to discuss the main effects only in the light

**TABLE 7.1**

Factorial Structure of Treatments

| | Factor 1: Cutting | |
|---|---|---|
| **Factor 2: Seeding** | **Level 1: Cutting Once/Year** | **Level 2: Cutting Twice/Year** |
| Level 1: with seeds | Treatment combination 1: F1 | Treatment combination 2: F2 |
| Level 2: no seeds | Treatment combination 3: NF1 | Treatment combination 4: NF2 |

of their interactions; it may not be possible to generalise about sowing seed because its effect on spider numbers depends on how frequently we cut the vegetation.

It might have been better to call our original treatments **treatment combinations**, as each treatment was a combination of two factors (Table 7.1).

Think of the treatment combination as the conditions that each replicate experiences. Some treatments may be the result of several "things" (**factors**) being applied. In our experiment we have four treatment combinations with a $2 \times 2$ factorial structure (because each of the factors is present at two levels).

To summarise:

- **Factor:** something which may have an effect (e.g., herbicide, moisture)
- **Level:** the state of a factor (e.g., present or absent; low, medium, or high)
- **Treatment:** a particular combination of one level of one factor plus one level of another factor
- **Replicates:** the number of individuals or plots that experience the same conditions

### 7.3.1 What Is "Hidden Replication"?

Another advantage of the factorial design is that of **hidden replication**. Because ANOVA can work out how much of the variation is due to which factor, the true number of replicates is all replicates with any one level of a factor. For example, although there are only four replicates in each treatment combination for the spider experiment, the total number of replicates for any level of a factor is eight (Table 7.2).

## 7.4 Getting More Out of the Analysis: Using the Factorial Design to Ask More Relevant Questions

Because there are three degrees of freedom for the treatment effect, we could ask up to three independent questions about the treatment combinations. It

**TABLE 7.2**

Hidden Replication in Factorial Design

| Seeding Level | Cutting Level | | Totals |
| --- | --- | --- | --- |
| | Cut Once/Year | Cut Twice/Year | |
| With flower seeds | 4 | 4 | Total replicates for with flower seed = 8 |
| Without flower seeds | 4 | 4 | Total replicates without flower seed = 8 |
| | Total replicates cut once/year = 8 | Total replicates cut twice/year = 8 | |

**TABLE 7.3**

Rearrangement of the Data to Show the Effects of Seed Sowing and Cutting Regime Separately

| | Cut Once | Cut Twice | Mean |
| --- | --- | --- | --- |
| Seed | 21,20,19,18 | 16,16,14,14 | 17.25 |
| No seed | 18,17,15,16 | 14,13,13,12 | 14.75 |
| Mean | 18.0 | 14.0 | 16.0 |

**TABLE 7.4**

Treatment Effects for Each Treatment and Expected Values for Each Plot

| | Cut Once | Cut Twice | Mean | Seed Effect |
| --- | --- | --- | --- | --- |
| Seed | 16 + 2 + 1.25 = 19.25 | 16 – 2 + 1.25 = 15.25 | 17.25 | (17.25 – 16) = +1.25 |
| No seed | 16 + 2 – 1.25 = 16.75 | 16 – 2 – 1.25 = 12.75 | 14.75 | (14.75 – 16) = –1.25 |
| Mean | 18.00 | 14.00 | 16 | (0) |
| Cut effect | (18 – 16) = +2 | (14 – 16) = –2 | (0) | |

is more interesting and efficient to ask questions about whether the factors differ rather than the treatment combinations.

To see this we rearrange the numbers from the data set so that we can see the effects of seed and cutting regime separately (Table 7.3). You can see that the mean number of spiders for plots that are cut once is 18, as opposed to 14 for those cut twice. Similarly, the mean for seeded plots is 17.25, compared with only 14.75 if there is no seed.

We can now work out the sums of squared deviations for the effect of cutting and seed from first principles quite easily (Table 7.4). As before, the effect of each level of a factor is the mean for all plots treated with that level minus the grand mean. Thus, the effect of sowing seed, over and above the grand mean, is *mean for all the seeded plots – grand mean*, which is 17.25 – 16 = +1.25. Similarly, the effect of no seed, over and above the grand mean, is mean for all seeded plots: grand mean is 14.75 – 16 = –1.25.

We can now start to fill in the ANOVA plan. We begin by putting in the total degrees of freedom and the total sum-of-squares; the data is the same as in previous analyses (see Section 6.6), so these will not have changed. In the top line we add the degrees of freedom for seed. The sum of squared deviations for seed is easy to work out; it is the square of the seed effect for each individual plot: $(+1.25)^2$ for all the seed plots and $(-1.25)^2$ for all the nonseed plots. There are 16 plots altogether, 8 of each, so the total is:

$$8 \times (+1.25)^2 + 8 \times (-1.25)^2 = 16 \times (1.25)^2 = 25$$

We can do an exactly analogous calculation with the cut effect (shown on the bottom row of Table 7.4). Thus, the effect of cutting only once over and above the grand mean is to add two spiders per plot, and the effect of cutting twice is to decrease the number by two spiders per plot on average. Again note that $+2 + (-2) = 0$, so there is one degree of freedom, and the sum-of-squares for the cutting regime is $16 \times 2^2 = 64$. As we already know the total degrees of freedom and sum-of-squares, we can fill in the error d.f. (13) and the error sum-of-squares (17) by subtraction and perform the remaining calculations much as we did for the blocked example in the last chapter:

```
ANOVA: Spiders vs. Seed, Cut

Factor    Type   Levels  Values
seed      fixed      2   seeded, unseeded
cut       fixed      2   once, twice

Analysis of variance for spiders

Source  DF        SS       MS      F       P
seed     1    25.000   25.000   19.12   0.001
cut      1    64.000   64.000   48.94   0.000
Error   13    17.000    1.308
Total   15   106.000
```

What we have now done is to test separately the hypotheses that flower seed makes a difference to spider numbers and that cutting regime makes a difference. Each of the tests is adjusted for the effects of the other, so we have managed to do two separate experiments with the same plots (very efficient). Note that the use of this technique depends on the fact that each combination of treatments has the same number of observations in it. If this turns out not to be true, we have to makes special arrangements, which we will discuss in Section 7.9. The model formula for this analysis is

$$spiders = seed + cut$$

As discussed at the beginning of Chapter 5, each term in the model represents the test of a different Null hypothesis. For each test in turn we ask whether the test statistic, F, is larger than expected if the associated Null hypothesis is true (Table 7.5), and we can see the quantities needed to make this judgment in the rows of the ANOVA table. Each of these tests is separate because each F-ratio takes into account the other term. We can say that seed

**TABLE 7.5**

The Different Terms in the Model and Their Associated Hypotheses

| Model Term | Research Hypothesis | Null Hypothesis |
|---|---|---|
| Seed | Spider numbers depend on whether plots have seed sown or not | Spider numbers are the same whether seed sown or not |
| Cut | Spider numbers depend on whether plots are cut once or twice a year | Spider numbers are the same whatever the cutting regime |

sowing has an effect ($F_{1,13} = 19.12$, $p = 0.001$) after allowing for the effect of cutting, and that cutting has an effect ($F_{1,13} = 48.94$, $p < 0.0005$) after allowing for the effect of seeding. It does not matter in which order we carry out the tests within this ANOVA.

## 7.5    Interactions

So far, we have assumed that each factor has an influence on spider numbers, irrespective of the level of the other factor. This need not be true; they could **interact**.

We relied on subtraction to get the error sum of squares in the last section. To get the error sum of squares directly, we need to work out the expected values for each cell. In general, it is *grandmean + seedeffect + cuteffect*. For the top left cell of Table 7.4, which represents plots that are seeded and cut once, it will be

$$grandmean + seedeffect + cuteffect = 16 + 1.25 + 2 = 19.25$$

In the body of Table 7.4 we have put the four expected values. Check to see if you can see where each comes from. Now, we know that, in general, the error sum-of-squares is the sum of the squared differences between each expected value and the corresponding observation, i.e., $\Sigma(observed - expected)^2$ for all the observations. Because the observed values have not changed, the difference between our two models must lie in the expected values. In the body of Table 7.6 we calculate the difference between the expected values derived from the original model, *spiders = treatment* (with the four treatment combinations considered separately) and the expected values for the new model, *spiders = seed + cut*. Notice that the expected values for the *spiders = treatment* model are the same as the means for each group.

### 7.5.1    Where Does the Interaction Term Come from?

If we think again about the model calculations in Table 7.4, we realise that there is a **constraint**. The *spiders = seeds + cut* model says that the difference between spider numbers on cut-once and cut-twice plots is $19.25 - 15.25 = 4$

**TABLE 7.6**

Discrepancies between Observed and Expected Values for the Unconstrained Model (*spiders=treatment*) and the Constrained Model (*spiders=cut+flowers*)

|  | Cut Once | Cut Twice |
|---|---|---|
| Seed | 19.5 – 19.25 = +0.25 | 15.0 – 15.25 = –0.25 |
| No seed | 16.5 – 16.75 = –0.25 | 13 – 12.75 = +0.25 |

for seeded plots and 16.75 – 12.75 = 4 for unseeded plots. The fact that these two differences are the same is not an accident; it is defined by the structure of the model, which says that the effect of cutting regime *adds on* to the effect of seeding. Thus, the same effect of cutting must be added regardless of which seeding treatment we are dealing with. However, when we look at the actual means for each group of plots, we find that the difference between cutting regimes is 19.5 – 15 = 4.5 for seeded plots, and 16.5 – 13 = 3.5 for unseeded plots. What is happening is that, in the original data, the effect of the cutting regime in unseeded plots was *slightly less* than it was in the seeded plots. This type of relationship is called a **statistical interaction,** and to capture it in our statistical model we must add an extra term, the **interaction term** *seeds*cut*.

### 7.5.2 Testing the Interaction: Is It Significant?

If we can calculate a sum-of-squares for the interaction term, then we can test it in the same way as we test any other effect.

We can calculate this because the interaction term is the squared deviations between a model with each factor represented separately (*spiders = seed + cut*) and a model that allows every treatment combination to have its own mean (in this case, *spiders = treatment*). We have the discrepancies we need in Table 7.6; the discrepancy is 0.25 for every observation, so the sum-of-squares for the interaction effect is $16 \times 0.25^2 = 1$. We can write this into the ANOVA plan, and we now find that the model part of the sum-of-squares adds up to 90, as it did in the model *spiders = treatment*, but the 90 has been split up into three components: the effect of seed, taking into account cutting; the effect of cutting, taking into account seed; and the interaction between the two, which is conventionally written Seed*Cut. (Table 7.6).

### 7.5.3 Degrees of Freedom for an Interaction Term

We now have the interaction sum-of-squares and an error sum-of-squares (Table 7.7). If we can work out the degrees of freedom for the interaction, we can calculate an F-ratio and p-value. We could just guess that as the sum of squares attributable to Seed, Cut and Seed*Cut is the same as that attributable to Treatment, then the degrees of freedom should be the same too, so

**TABLE 7.7**

Degrees of Freedom

| Source | Degrees of Freedom | Sum of Squares | Mean Square | F-Ratio |
|--------|--------------------|----------------|-------------|---------|
| Seed | 1 | 25 | | |
| Cut | 1 | 64 | | |
| Seed*Cut | 1 | 1 | | |
| Error | 12 | 16 | | |
| Total | 15 | 106 | | |

we can deduce that there is one degree of freedom for the interaction term, giving three altogether for the predictive part of the model. We can be a bit clearer about this, though, by looking at Table 7.6 again and realizing that if we know the top left discrepancy is +.25, then we can fill in all the others as shown without doing any more calculations (Sudoku players will have no difficulty with this idea, but it takes a few minutes for the rest of us to see it). Thus, there is only one independent piece of information here, and hence one degree of freedom.

The general rule for the degrees of freedom of an interaction A*B is that it is d.f. (A) × d.f. (B). So if A had 2 levels and B had 3, the d.f. for the interaction is $(2 - 1) \times (3 - 1) = 2$.

When MINITAB does the calculations, we get:

```
ANOVA: Spiders vs. Seed, Cut

Factor   Type   Levels   Values
seed     fixed      2    seeded, unseeded
cut      fixed      2    once, twice

Analysis of variance for spiders

Source    DF      SS       MS      F       P
seed       1   25.000   25.000   18.75   0.001
cut        1   64.000   64.000   48.00   0.000
seed*cut   1    1.000    1.000    0.75   0.403
Error     12   16.000    1.333
```

There are now three hypothesis tests to consider (Table 7.8).

In this case, the interaction term is small, and the corresponding hypothesis test gives a p-value of 0.403, well above the normal rejection criterion. We can therefore conclude that the effect of flower seed is probably the same whether the plots are cut once or twice a season, and the effect of cutting is the same whether the plots are seeded or not; in fact, the effects of the two factors are **additive**. According to the model, the effect of flower seed *adds together* with the effect of cutting regime to determine spider numbers.

Although interaction terms are listed after the main effects in the ANOVA table the components of the model must be tested in the order of decreasing complexity. If the Null hypothesis that there is no interaction (the effects of

**TABLE 7.8**

Hypotheses Associated with Terms in the Model

| Model Term | Research Hypothesis | Null Hypothesis |
|---|---|---|
| Seed*cut | The effect of seeds depends on whether the plots are cut once or twice (and *vice versa*) | The effects of the two are independent of one another |
| Cut | Spider numbers depend on whether plots are cut once or twice a year | Spider numbers are the same whatever the cutting regime |
| Seed | Spider numbers depend on whether plots have seed sown or not | Spider numbers are the same whether seed sown or not |

the two are independent) is rejected, i.e., if there *is* an interaction, then both seed and cut *must both be having important effects even if either the cut or the seed term appear not to be significant.* In this situation, the p-values for the "main" effects are not meaningful criteria on which to accept or reject the hypothesis that the seeding or cutting treatments have no effect when acting alone. We can say the main effects must be important because the interaction is significant.

## 7.6 Adding Blocking to the Factorial Analysis

No new concept is involved here; we just add in the block effect, which is the sum of squared deviations of the block means about the grand mean exactly as it was when we first met it in Chapter 6. Thus, the block sum of squares and degrees of freedom will be the same as before:

```
Factor   Type    Levels   Values
block    fixed        4    1 2 3 4
cut      fixed        2    once twice
seeds    fixed        2    seeded unseeded

Analysis of variance for spiders

Source      DF       SS       MS        F       P
block        3   13.500    4.500    16.20   0.001
cut          1   64.000   64.000   230.40   0.000
seeds        1   25.000   25.000    90.00   0.000
cut*seeds    1    1.000    1.000     3.60   0.090
Error        9    2.500    0.278
Total       15  106.000
```

Block is significant ($F_{3,9} = 16.2$, $p = 0.001$). We can look at block first because it is not involved in any interaction. For the other terms the main difference to note is that because the error sum-of-squares has been reduced by 13.5, the error mean square is now only $2.5/9 = 0.278$. Consequently, the F-ratios for cut,

seeds, and cut*seeds are all larger than they were before, although the interpretation does not change in this case. We still find that the interaction term
is not significant ($F_{1,9} = 3.60$, $p = 0.09$), so we do not reject the Null hypothesis
that the factors are additive. We can, therefore, look at the p-values for the
main effects. In this case, they are both highly significant, as in the unblocked
analysis.

We can now look at what the actual effects are. For example, do you get
more spiders if you sow wildflower seed? We can display the means of each
factor level by asking for them when we do the analysis. This function should
be available in all programs, but in MINITAB, it is by going into RESULTS
and specifying the model terms for which you want the means. In this case,
they are: block cut seeds cut*seeds.

```
Means

block  N   spiders
    1  4    17.250
    2  4    16.500
    3  4    15.250
    4  4    15.000

seeds       N   spiders
seeded      8    17.250
unseeded    8    14.750

cut     N   spiders
once    8    18.000
twice   8    14.000

seeds         cut  N   spiders
seeded       once  4    19.500
seeded      twice  4    15.000
unseeded     once  4    16.500
unseeded    twice  4    13.000
```

We can plot the means in order to visualise the effects of the different
factors (Figure 7.1 and Figure 7.2).

---

## 7.7  How to Interpret Interaction Plots: The Plant Hormone Experiment

We can plot the interaction from the means given in the output or in
MINITAB as a separate plot **(Stat>ANOVA>Interactions plot, Responses**:
spiders, **Factors**: seeds cut). The interactions plot for spiders is shown in
Figure 7.3. The fact that the lines are nearly parallel reflects the fact that the

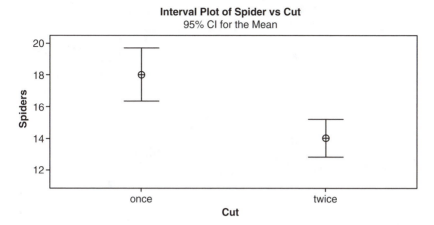

**FIGURE 7.1**
Plot showing the effect of cutting on spider numbers. Circles show the mean, and bars represent the 95% confidence level.

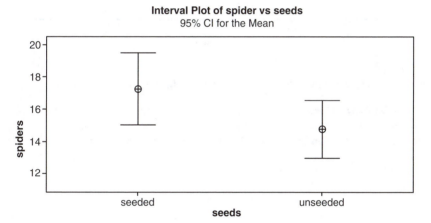

**FIGURE 7.2**
Graph to show the effect of sowing wildflowers on the mean spider number. Bars represent the 95% confidence interval.

interaction was not significant, so we will use another example to demonstrate the meaning of a significant interaction term.

We have designed an experiment to investigate the importance of increasing amounts of phosphate (1, 2, 3, and 4 mg per culture) and of contact with living plant tissue (present or absent) on the extension of a bacterial colony (millimeter per day).

The experiment was carried out by inoculating a carbohydrate medium on petri dishes with the bacterium. One set of each of the 8 treatments was established on each of 3 days, making 24 dishes in all. The amounts of extension growth are analysed in MINITAB in Table 7.9.

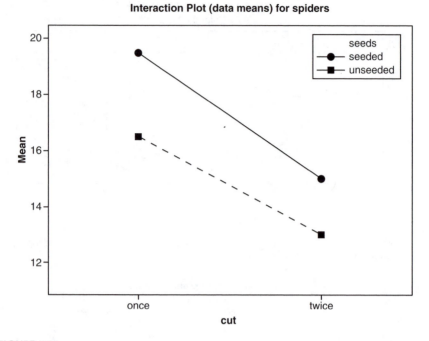

**FIGURE 7.3**
Interaction plot showing the effect of cutting and sowing on spider numbers.

**Stat>ANOVA>Balanced ANOVA Reponses:** extension
**Model:** experiment+contact+phosphate+contact*phosphate
**ANOVA: Extension vs. Experiment, Contact, Phosphate**

```
Factor        Type  Levels  Values
experiment    fixed      3  1, 2, 3
contact       fixed      2  contact, no contact
phosphate     fixed      4  1, 2, 3, 4

Analysis of variance for extension

Source               DF      SS       MS       F      P
experiment            2   4.000    2.000    0.41  0.670
contact               1   2.667    2.667    0.55  0.471
phosphate             3 166.000   55.333   11.39  0.000
contact*phosphate     3  57.333   19.111    3.93  0.031
Error                14  68.000    4.857
Total                23 298.000
```

**TABLE 7.9**

Bacterial Growth (Extension in millimeter per day) for Eight Treatment Combinations

| Row | Experiment | Contact | Phosphate | Extension | Treatment |
|-----|-----------|---------|-----------|-----------|-----------|
| 1 | 1 | Contact | 1 | 10 | 1 |
| 2 | 1 | No contact | 1 | 6 | 2 |
| 3 | 1 | Contact | 2 | 13 | 3 |
| 4 | 1 | No contact | 2 | 11 | 4 |
| 5 | 1 | Contact | 3 | 14 | 5 |
| 6 | 1 | No contact | 3 | 20 | 6 |
| 7 | 1 | Contact | 4 | 16 | 7 |
| 8 | 1 | No contact | 4 | 22 | 8 |
| 9 | 2 | Contact | 1 | 12 | 1 |
| 10 | 2 | No contact | 1 | 10 | 2 |
| 11 | 2 | Contact | 2 | 13 | 3 |
| 12 | 2 | No contact | 2 | 13 | 4 |
| 13 | 2 | Contact | 3 | 14 | 5 |
| 14 | 2 | No contact | 3 | 14 | 6 |
| 15 | 2 | Contact | 4 | 14 | 7 |
| 16 | 2 | No contact | 4 | 18 | 8 |
| 17 | 3 | Contact | 1 | 14 | 1 |
| 18 | 3 | No contact | 1 | 10 | 2 |
| 19 | 3 | Contact | 2 | 12 | 3 |
| 20 | 3 | No contact | 2 | 10 | 4 |
| 21 | 3 | Contact | 3 | 10 | 5 |
| 22 | 3 | No contact | 3 | 14 | 6 |
| 23 | 3 | Contact | 4 | 16 | 7 |
| 24 | 3 | No contact | 4 | 18 | 8 |

The significant interaction shows that the main effects of contact and phosphate must be important and that the action of each depends on the level of the other (contact*phosphate $F_{3,14} = 3.93$, $p = 0.031$).

The bacteria not in contact with plant tissue show a great response to increasing amounts of phosphate. In contrast, the bacteria that are in contact with plant tissue show much less response. Perhaps this is better displayed in an interaction diagram (Figure 7.4). We might speculate on the biological reasons for such a difference.

An interaction means that the effect of one factor depends on the level of the other. Therefore, when we plot the effects, the lines should not be parallel. Figure 7.4 shows the interaction plot for the effect of phosphate and contact with neighbouring colonies on the growth of bacteria. These lines are clearly not parallel, and in the analysis the interaction was significant. Contrast this with the near-parallel lines in Figure 7.3, where the interaction term was not significant.

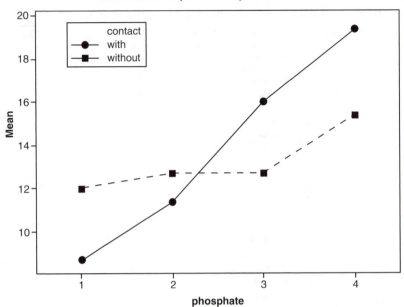

**FIGURE 7.4**
Interaction plot showing the effect of phosphate and contact on bacterial growth (millimeters per day).

## 7.8   Loss of Data and Unbalanced Experiments

Almost every biologist has a story about losing data or replicates. These range from "someone cleared out the freezer and threw away half of my cockroaches" to "two of my plots were bombed by the government and I had to leave."

Missing data are common, but we still want to analyse the remaining data if possible. When you enter the data into your statistics package, it is important to find out what it will recognize as a missing value. MINITAB uses the "*" character, but other packages use "." or even just a blank cell. Zero is not a good choice because it will be confused with a real value of zero. Leaving the data out altogether is not always possible, as you might have some of the measurements for a damaged plot but not others.

Because you have fewer observations altogether, you have less information. As a result, the experiment is less likely to be able to detect differences between the populations you are comparing. A more difficult problem is that you now have more information about some treatments than about others. If we confine ourselves to a one-way analysis of variance, this is not too bad, and the analysis can be carried out in MINITAB as usual. It takes into account the differing replication between the treatments through the calculations of the standard errors for the group means, but some caution is still needed in interpreting the hypothesis test. However, when we move to a two-factor model, such as *spiders = cut + flowers,* things become more difficult.

### 7.8.1   An Example of Loss of Data: The Plant Hormone Experiment

In a wholly randomised experiment, we wish to compare the effects of six different plant hormones on floral development in peas. A standard amount was applied to the first true leaf of each of four plants for each hormone, making 24 plants in the experiment. Unfortunately, two plants were knocked off the greenhouse bench and severely damaged: one from hormone 2 and one from hormone 6 (Table 7.10). So we have data on the number of the node (the name given to places on the stem where leaves or flowers may appear) at which flowering first occurred for only 22 plants. Still, a one-way ANOVA can analyse this without any problem, as there are replicates in each treatment.

**ANOVA: node versus hormone**

```
Factor    Type    Levels  Values
hormone   fixed       6   1, 2, 3, 4, 5, 6

Analysis of Variance for node

Source    DF       SS      MS      F      P
hormone    5  101.008  20.202   4.61  0.009
Error     16   70.083   4.380
Total     21  171.091

S = 2.09289   R-Sq = 59.04%   R-Sq(adj) = 46.24%

Means

hormone   N     node
1         4   50.000
2         3   54.667
3         4   50.000
4         4   53.750
5         4   53.000
6         3   55.667
```

**TABLE 7.10**

Node Number per Plant in a Plant Hormone Experiment with a Wholly Randomised Design

|          | Hormone | | | | | |
| :------: | :---: | :---: | :---: | :---: | :---: | :---: |
| Replicate | 1 | 2 | 3 | 4 | 5 | 6 |
| 1 | 49 | 53 | 49 | 53 | 51 | 58 |
| 2 | 52 | * | 50 | 57 | 56 | * |
| 3 | 51 | 54 | 52 | 54 | 51 | 54 |
| 4 | 48 | 57 | 49 | 51 | 54 | 55 |

**TABLE 7.11**

Node Number per Plant in a Plant Hormone Experiment as a Randomised Complete Block Design

|          | Hormone | | | | | |
| :------: | :---: | :---: | :---: | :---: | :---: | :---: |
| Block | 1 | 2 | 3 | 4 | 5 | 6 |
| 1 | 49 | 53 | 49 | 53 | 51 | 58 |
| 2 | 52 | * | 50 | 57 | 56 | * |
| 3 | 51 | 54 | 52 | 54 | 51 | 54 |
| 4 | 48 | 57 | 49 | 51 | 54 | 55 |

## 7.8.2 Calculating Standard Errors Where You Have Missing Observations

ANOVA provides strong evidence to reject the Null hypothesis that all six hormones affect floral initiation in similar fashion. If we wish to calculate standard errors for treatment means, we must remember that they are based on different numbers of replicates. Remember that the formula is

$$SEmean = \sqrt{\frac{ErrorMeanSquare}{n}}$$

So, for hormones 2 and 6, where we have only three replicates, the SE for each mean is the square root of $4.38/3 = 1.46$, whereas, for the remaining hormones, it is the square root of $4.38/4 = 1.095$, which is considerably smaller.

This seems to be satisfactory, but in fact we realised that our glasshouse was not a homogeneous environment and so laid out our experiment in four randomised complete blocks. The one-way analysis is thus not really valid because the randomisation of the pot positions was carried within blocks and not across the whole greenhouse. We must "analyse as we randomise," and so, we now have a two-way analysis of variance. However, this is not balanced (Table 7.11).

If we want to compare the mean of hormone 2 with that of hormone 4, we have a problem. It may differ because the hormone has a different effect, but it may also differ because hormone 2 was not represented in block 2, which could be a location in the glasshouse that is especially good, or especially bad, for plant growth.

---

## 7.9 Limitations of ANOVA and the General Linear Model (GLM)

If we try to analyse the data using the usual ANOVA command in MINITAB, we would get an error message:

```
* ERROR * Unbalanced design. A cross tabulation of your factors
        * will show where the unbalance exists.
```

This is because we do not have one representative of each treatment in each block. Remember, we said that factorial designs have to have replicates of every treatment combination. By including block, the missing values now become critical. To analyse these data, we need to use a more flexible method. An extension of ANOVA, which can handle unbalanced designs (as well as a large number of other complications), is the General Linear Model (GLM). To fully understand this, you need to read a text which covers it in detail (e.g., Grafen and Hails, 2002). We just dip our toes in here, so see how the simple case of missing data leading to unbalance will look.

MINITAB implements GLM using **Stat>Anova>General Linear Model** (not available in the Student edition, unfortunately). If this is used in place of **Stat>Anova>Balanced Anova**, an analysis is possible:

```
General Linear Model: Node vs. Block, Hormone

Factor     Type   Levels  Values
Block      fixed      4   1, 2, 3, 4
Hormone    fixed      6   1, 2, 3, 4, 5, 6

Analysis of variance for node, using adjusted SS for tests:

Source   DF   Seq SS   Adj SS   Adj MS     F      P
Block     3    6.841   23.465    7.822   2.18  0.139
Hormone   5  117.632  117.632   23.526   6.56  0.003
Error    13   46.618   46.618    3.586
Total    21  171.091
```

The main thing to notice about this output is the two columns of sums of squares. These are "sequential" (Seq SS) (also known as "Type I") and "adjusted" (Adj SS) (also known as "Type III"). When an experiment is balanced, the order in which we add the terms to the model (here, hormones and blocks) does not affect the outcome. There is only one value for the sum-of-squares due to hormones, whether we ask it to be fitted before or after blocks. However, in an unbalanced experiment, the order in which the terms are fitted is very important. The sequential (Type I) sum-of-squares in the example asks the question "How much variance does block explain?" and "How much variance does hormone explain when block has already been put into the model?" because block has been fitted first and then hormone. This was the order in which we placed the terms in our model formula. However, in the adjusted (Type III) sum-of-squares column, each term has the sum of squares appropriate to it if it were fitted last in the model. So, here, if block is fitted before hormones, its SS is 6.841, but if it is fitted after blocks, its SS is 23.465. It would be sensible to use the adjusted sum-of-squares here because we want to know the effect of each term adjusted for the other, as we did for the spiders analysis in Chapter 6. Here, it represents evidence for differences between the hormone effects after taking into account the fact that some hormones were not represented in all of the blocks. We would be unlikely, in reality, to be interested in the block effect before the effects of hormone were taken into account, but if we had a two-factor model such as *spiders = cut + seed*, then it would become important.

The GLM is much more than simply a method for dealing with missing data. It allows us to combine much of what has been covered in Chapter 3 to Chapter 7 in a single framework, as well as dealing with predictors that are continuous rather than categories. The conventional method for dealing with continuous predictors (Regression) is the subject of the next chapter. The GLM approach is treated in detail by Grafen and Hails (2002).

# 8

## Relating One Variable to Another

> Statistics have shown that mortality increases perceptibly in the military during wartime.

> —**Robert Boynton**

In the previous three chapters we have been concerned with the relationship between a single continuous variable and one or more categorical variables. Thus, in Chapter 5 we asked whether the number of spiders in a quadrat (continuous) differed depending on whether we sowed wildflower seed (categorical). In this chapter we consider what to do when we have two quantitative variables that we think might be related to one another.

## 8.1 Correlation

The simplest question we could ask about two continuous variables is whether they vary in a related way, i.e., is there a **correlation** between them? For example, the concentration (ppm) of two chemicals in the blood might be measured from a random sample of 14 patients suffering to various extents from a particular disease. If a consequence of the disease is that both chemicals are affected, we should expect patients with high values of one to have high values of the other and *vice versa*. Table 8.1 shows the concentrations of chemical A and of chemical B in the blood of 14 such patients.

The data are shown as a scatter plot in Figure 8.1a. For comparison, Figure 8.1b shows the same data but with the column for B scrambled into a random order. In the graph of the "real" relationship, you can see that generally low concentrations of A tend to be associated with low concentrations of B, giving a "bottom left to top right" look to the graph. If we break up the relationship between each patients' A and B concentration by randomizing column B, then the pattern disappears (Figure 8.1b).

How do we characterize this relationship? Bearing in mind that what we are claiming is that *relatively* large concentrations of A are associated with

**TABLE 8.1**

Concentrations of Chemicals A and B in the Blood of 14 Patients with z-Scores

| Row | A | B | z-ScoreA | z-ScoreB | z-ScoreA * z-ScoreB |
|---|---|---|---|---|---|
| 1 | 23.6 | 15.2 | 2.04881 | 2.08005 | 4.26163 |
| 2 | 23.7 | 12.3 | 2.06308 | 1.26253 | 2.60470 |
| 3 | 7.0 | 10.9 | 0.32006 | 0.86786 | 0.27777 |
| 4 | 12.3 | 10.8 | 0.43626 | 0.83967 | 0.36632 |
| 5 | 14.2 | 9.9 | 0.70740 | 0.58596 | 0.41451 |
| 6 | 7.4 | 8.3 | 0.26298 | 0.13491 | 0.03548 |
| 7 | 3.0 | 7.2 | 0.89088 | 0.17518 | 0.15607 |
| 8 | 7.2 | 6.6 | 0.29152 | 0.34433 | 0.10038 |
| 9 | 10.6 | 5.8 | 0.19367 | 0.56985 | 0.11036 |
| 10 | 3.7 | 5.7 | 0.79098 | 0.59804 | 0.47304 |
| 11 | 3.4 | 5.6 | 0.83379 | 0.62623 | 0.52215 |
| 12 | 4.3 | 4.2 | 0.70536 | 1.02090 | 0.72010 |
| 13 | 3.6 | 3.9 | 0.80525 | 1.10547 | 0.89018 |
| 14 | 5.4 | 3.1 | 0.54839 | 1.33099 | 0.72990 |
| Mean | 9.24 | 7.81 | | Sum | 10.82 |
| S.D. | 7.01 | 3.55 | | Sum/(n1) | 0.832 |

*relatively* large concentrations of B, it makes sense to express each measurement as a z-score, which is defined for each value of x as

$$zscore(x) = \frac{(x - \bar{x})}{sd(x)}$$

just as we saw in Chapter 2. In Table 8.1 we have calculated the z-scores for concentrations of A and B, and these are plotted in Figure 8.2. You can see straightaway from the figure that most of the points fall in the top right or the bottom left quadrants made by the x- and y-axes, so that in terms of z-scores, a negative zscoreA tend to be associated with a negative zscoreB, and positive scores also tend to fall together. The three rows where this is not the case are the three rows corresponding to the points in the top left and bottom right quadrants. If there was no association between A and B, we would find that rows consisting of + +, − +, + −, and − − z-scores would occur with about equal frequency so that each quadrant had about the same number of points in it.

So, if we just multiply the two z-scores together for each point and add up the results, we get a number that will tend to be large and positive if all the scores are in the bottom left or top right quadrants but which would tend to be close to zero if the points were evenly distributed. Unfortunately, the sum of products of the z-scores gets bigger and bigger the more rows there are, so we standardise it by dividing by n − 1, where n is the number of pairs of data points. This should mean that the value obtained is directly comparable between studies with different numbers of data points.

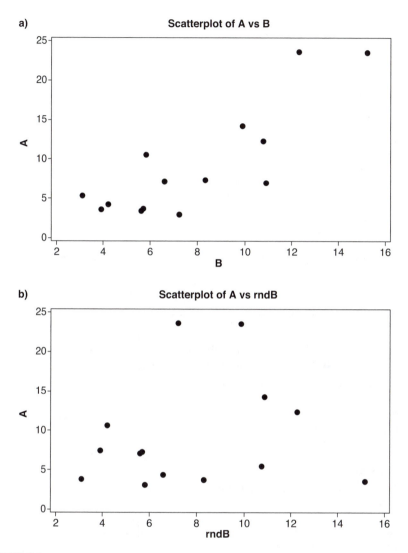

**FIGURE 8.1**

(a) A positive relationship between B and A. (b) Column B has been randomised, and now there is no relationship between A and B.

The result of these calculations is a new test statistic called **Pearson's correlation coefficient**, written *r* for short. It is also sometimes known as the **product moment correlation coefficient**. The test statistic *r* can be positive or negative but always lies between −1 and +1 (Box 8.1). A value near +1 indicates a strong positive correlation, whereas a value near −1 shows a strong negative relationship. A value of 0 may indicate a lack of relationship, though it could also mean that a more complicated relationship is present

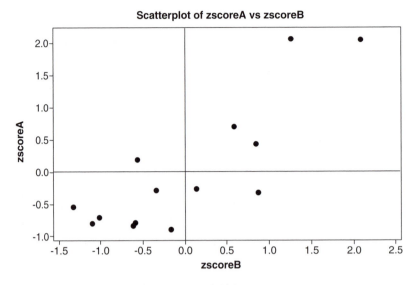

**FIGURE 8.2**
Scatter plot of the z-score of A against the z-score of B.

(a curve, for example). MINITAB will calculate the correlation coefficient for us. **Stat>Basic Statistics>Correlation**. Notice that it does not matter which way round we specify A and B; the relationship is symmetrical.

```
Correlations: A, B
Pearson's correlation of A and B = 0.832
p-Value = 0.000

Correlations: B, A
Pearson's correlation of B and A = 0.832
p-Value = 0.000
```

Like other test statistics, $r$ has a known distribution if the relevant Null hypothesis ($H_0$) is true. In this case $H_0$ is that there is no association between the measurements, so the expected correlation coefficient is zero. We can find out whether the calculated value of Pearson's $r$ indicates a significant correlation by comparing it with critical values found in a statistical table (Appendix C, Table C.8). If our calculated value (ignoring the sign, as it is only the size of the coefficient rather than its sign that is important for the hypothesis test) is greater than the value found in the table for the appropriate degrees of freedom n, and $p = 0.05$, we conclude that there is a significant correlation. In this case, the critical value for $p = 0.05$ and $n - 2 = 12$ d.f., whereas for $p = 0.01$ it is 0.661. So, we have strong evidence for a significant correlation. MINITAB has printed the actual p-value ($p < 0.001$).

## 8.2 Calculating the Correlation Coefficient, and a New Idea: Covariance

The calculations for $r$ in Table 8.1 can be expressed by the following equation:

$$r = \frac{1}{(n-1)} \sum zscore(x) \times zscore(y) = \frac{1}{n-1} \sum \frac{(x-\bar{x})}{sd(x)} \times \frac{(y-\bar{y})}{sd(y)}$$

However, we can simplify this by taking the terms involving $sd(x)$ and $sd(y)$ outside the summation term; they will be the same for every pair of $x$, $y$ values. (Mathematicians call this move *factorisation* — nothing to do with statistical factors!). This gives a simpler formula, which is what we normally use when calculating $r$ by hand:

$$r = \frac{\sum (x-\bar{x}) \times (y-\bar{y})}{n-1} \times \frac{1}{sd(x) \times sd(y)}$$

You can see here that a new quantity

$$\frac{\sum (x-\bar{x}) \times (y-\bar{y})}{n-1}$$

has emerged. This looks suspiciously like a variance, except that, instead of squaring the deviations of $x$ values about their mean, or $y$ values about theirs, we multiply the deviations together and then scale by $n-1$. This quantity is called the **covariance** and is the fundamental quantity in dealing with relationships between continuous variables. If we write

$$cov(x,y) = \frac{\sum (x-\bar{x}) \times (y-\bar{y})}{n-1}$$

then we have the conventional calculation for Pearson's $r$:

$$r = \frac{cov(x,y)}{sd(x) \times sd(y)}$$

MINITAB will calculate the covariance for us:

```
Covariances: A, B

          A          B
A   49.1057
B   20.6805    12.5834
```

The diagonal terms are the variances for A and B, which are just the squares of standard deviations given in Table 8.1. So we have

$$r = \frac{20.6805}{\sqrt{49.1057} \times \sqrt{12.5834}} = \frac{20.68}{7.01 \times 3.55} = 0.83.$$

---

## BOX 8.1   Why Can *r* Never Be More Than 1?

The most extremely close correlation we could think of would be the correlation of a measurement with itself. The correlation of x with x would be

$$r = \frac{cov(x,x)}{sd(x) \times sd(x)} = \frac{\sum (x - \bar{x}) \times (x - \bar{x}) \Big/ (n-1)}{sd(x)^2} = \frac{var(x)}{sd(x)^2} = 1$$

MINITAB confirms that this is the correct result:

```
Correlations: A, A
Pearson's correlation of A and A = 1.000
p-Value = *
```

---

## 8.3   Regression

Correlation is, as we have seen, symmetrical. Neither of the variables concerned is either the response or the predictor, in the sense that the statistical model underlying ANOVA defines them. For example, in Chapter 6, we considered the effect of four different combinations of grass cutting and seeding on the number of spiders in field margins. We tested the Null hypothesis that all treatments (the predictors) have the same effect on spider numbers (the response). The treatments in that analysis had no particular natural ordering, and the outcome of the hypothesis test does not depend on whether we consider (for example) cutting twice with flower seed is "more or less" than cutting once with no seed. However, we often find that the predictors consist of increasing amounts of a factor.

As a simple case, we will look at the temperatures measured at a weather station in Alaska over a period of 12 years. The mean daily maximum temperature for July was measured every 3 years from 1979 to 1991. Here the treatment consists of going back to the weather station every 3 years to get another set of measurements. The usual practice is to put the **response variable** (temperature) on the left hand or y-axis, and the predictor (year) on the horizontal or x-axis. The response variable is also known as the **dependent variable** because it may depend on (be affected by) the magnitude of the x-variable (the **independent variable).**

We have already seen that, if there is no relationship between the two variables, the plot of response against predictor will show a random scatter of points (Figure 8.1b). If we found this, we would not need to carry out any further analysis, as knowing the position on the x-axis does not help us predict the size of the response. However, if we plot the data as collected, we do see a pattern. The points seem to lie about a straight line. This underlines the fact that the ordering of the years is an important feature of the data in a way that was not the case in the spiders analysis in Chapter 5 to Chapter 7. Here the line slopes up, so we can say that there appears to be a *positive* correlation (Figure 8.3). The longer after 1979 it is, the higher the mean July temperature; if it had sloped down, we would say there was a negative relationship in which more of x leads to less of y.

The correlation coefficient allows us to say whether there is a relationship or not, but we need to find a way of describing the extent to which the variation in y (temperature) can be explained by the variation in x (year). In terms of a statistical model, we are looking for a relationship of the form

$$temperature = year.$$

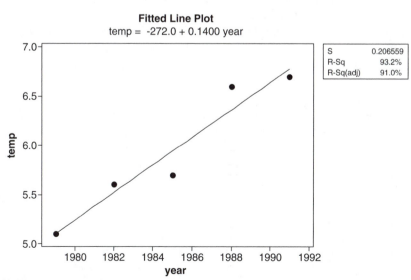

**FIGURE 8.3**
Fitted line plot of mean July temperature against year.

First, we will consider how to test whether our results provide evidence for a significant straight-line (linear) relationship or not. The process used is to find the **best-fit straight line**. As with many other statistical tests, we will be estimating parameters, in this case those that describe the line, and testing a hypothesis, in this case whether the positive or negative slope is more than would be expected if there was, in fact, no such relationship. The name for this procedure is **linear regression**.

## 8.4   Linear Regression

Sir Francis Galton (1822–1911) thought that the mean height of children within a family should tend to equal the mean height of their parents. In other words, he thought that tall parents would produce tall children. What he found was that the mean height of the children tended to be closer to the population mean than the mean height of the parents; he described this as "regression" (going back) towards the mean population height. The name **regression** has stuck since to describe the relationship that exists between two or more quantitative variables (in Galton's case, parents' height and height of child; in our example, temperature and year).

In Table 8.2 we show the data, and the calculations of the basic quantities needed to find the variance in temperature over the 12 years, and the variance in times.

Thus, the variance of temperatures is the sum of squared deviations of the temperatures about the mean temperature (5.94), squared (column 7), summed up and divided by $n-1$ in the usual way to give 0.473. A similar set of calculations with the deviations of year about the mean year (1985, which may seem a slightly odd idea but bear with us) gives the variance in the predictor quantity of 22.5. By themselves, these two measures do not

**TABLE 8.2**

Data Collected at Weather Station in Alaska with the Basic Quantities Required to Calculate Variances of Temperature °C and Year

| Temperature $y$ | Year $x$ | (3) $y-\bar{y}$ | (4) $x-\bar{x}$ | (5) $(y-\bar{y})(x-\bar{x})$ | (6) $(x-\bar{x})^2$ | (7) $(y-\bar{y})^2$ |
|---|---|---|---|---|---|---|
| 5.1 | 1979 | −0.84 | −6 | 5.04 | 36 | 0.7056 |
| 5.6 | 1982 | −0.34 | −3 | 1.02 | 9 | 0.1156 |
| 5.7 | 1985 | −0.24 | 0 | 0.00 | 0 | 0.0576 |
| 6.6 | 1988 | 0.66 | 3 | 1.98 | 9 | 0.4356 |
| 6.7 | 1991 | 0.76 | 6 | 4.56 | 36 | 0.5776 |
| Mean $\bar{y}=5.94$ | Mean $\bar{x}=1985$ | | Sums: | SPXY = 12.6 | SSX = 90 | SSY = 1.892 |
| | | | Divide by (n − 1) | 12.6/4 | 90/4 | 1.892/4 |
| | | | To get: | Cov(x,y)3.15 | Var(x) 22.5 | Var(y) 0.473 |

give us any information about how the temperature and year information are related to each other, but it turns out that the **covariance** that we met in Section 8.2 is just what we need, and this is calculated in column 5.

## 8.5   The Model

Like ANOVA (Chapter 6), the basis for regression analysis is a statistical model. The basic idea is the same as before: we are going to try to account for the variation in temperature in terms of things we know about (the passage of time) and things we do not (everything else). In ANOVA, we would write the equation for the expected temperature as

$$temperature = grandmean + time\_effect$$

However, we are proposing that the relationship is a straight line relating July mean temperature to year, so it makes better sense to write it as the equation for a straight line (see Box 8.2 for more explanation). The expected temperature is therefore

$$temperature = const + slope*time$$

where *const* is a constant, i.e., a number, and *slope* is the slope of the best-fitting line.

## Box 8.2 Describing a Straight Line by an Equation

First we put a y-axis and an x-axis on our graph. We always plot the *response* or *dependent* variable on the y-axis, and the *predictor* or *independent* variable on the x-axis. It is very important to get this the right way around.

To specify a particular straight line, we need to know two things:

- The value of y when x is zero (where the line cuts the y axis, known as the *intercept*)
- How much the line goes up (or down) for each increase of one unit in x (the *slope* or *gradient* of the line)

You may already have met the equation of a straight line in mathematics as

$$y = mx + c.$$

In this form, the letter $c$ represents the intercept, the value of $y$ when $x$ is zero, and the letter $m$ represents the slope.

In statistics, it is standard practice to use the letter *a* to represent the intercept and the letter *b* to represent the slope. The equation is thus written:

$$y = a + bx.$$

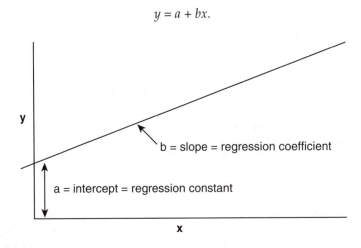

**FIGURE 8.4**
The slope and intercept of a regression line.

See Figure 8.4. We write the equation this way around because it is easily extended to more complicated models like multiple regression, in which, for example, we might wish to try to explain changes in temperature by knowing both the year and (say) the latitude of several weather stations for which we have data. It also underlines the analogy with the ANOVA:

$$response = grandmean + treatment\_effect.$$

First, imagine that we have the five July mean temperatures but we do not yet know which one comes from which year. The sum-of-squares of the differences between the temperature observations and their overall mean represents the total amount of variation in our response variable (SSY in Table 8.2).

We now want to divide that variation into two parts: the amount attributable to the passage of time (year) and the rest (random variation). This is similar to dividing up the variation in ANOVA into treatment and error effects, but we ask the question in a slightly different way. Specifically, if the plot of the data shows an approximately straight-line relationship with a slope, how much of the total variation in temperature (*y*) can be explained by a linear relationship with year (*x*)?

If the real relationship is close, we will find that almost all of the variation in *y* can be accounted for by knowing *x*. In this case, the best-fit line (the line

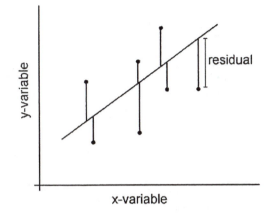

**FIGURE 8.5**
Residuals plotted above and below the line.

that best describes the relationship) will be close to all the points on the graph as we see in Figure 8.3. If there are many other factors that may affect the July mean temperature (for example, average cloud cover during the month), then the scatter of points about the line will be much wider.

Each year we noted the observed value of the mean July temperature $y$. The line of best fit calculated to describe this relationship will enable us in future to read off an **expected** or **fitted** or **predicted** value of $y$ for any particular value of $x$. This is unlikely to be identical to the observed value of $y$ for that value of $x$. The difference between the two is called a residual (Figure 8.5), just as with ANOVA. It tells us the discrepancy between our model (the straight line) and the data.

$$residual = observed\ y - fitted\ y$$

A residual may be a positive or negative number. The method used to obtain the line of best fit is to minimise the sum-of-squares of the distances of the observations from the line in a *vertical* direction. Another way of saying this is to minimise the sum-of-squares of the residuals. This is known as a **least-squares fit**. The vertical direction is used because we are trying to account for variation in $y$, which is on the vertical axis.

For reasons beyond the scope of this book (for a proof see Samuels and Witmer (2003) Appendix 12.1), it turns out that the best-fit line has the following properties:

- Its slope is given by

$$\frac{cov(x, y)}{var(x)}$$

Using the quantities calculated in Table 8.1, this give us

$$\frac{3.15}{22.5} = 0.14 \text{ degrees per year}$$

- It must pass through the point $\bar{x},\bar{y}$. This means that once we know the slope, we can get the constant by a simple bit of algebraic manipulation of the equation

$$\bar{y} = const + slope \times \bar{x}$$

$$\text{So } \bar{y} - slope \times \bar{x} = const = 5.94 - 0.14 \times 1985 = -271.96$$

Thus, the line of best fit for this data is *temperature* = −271.96 + 0.14 × *year*. Let us ask MINITAB to carry out the analysis for our weather station data:

```
Data Display
Row   Year   Temp
  1   1979   5.1
  2   1982   5.6
  3   1985   5.7
  4   1988   6.6
  5   1991   6.7
```

If we were doing the analysis by hand, we would plot the data first and, if the relationship looked plausible, proceed with the (rather tedious) calculations needed. If we use a computer, the obvious thing to do is to use the "fit line" procedure. This plots the data, carries out the statistical analysis, and puts the best-fit line on the graph in a single operation. The graph is shown in Figure 8.3, and the analysis is in the output that follows:

```
Regression Analysis: Temperature vs. Year

The regression equation is
temp = -272.0 + 0.1400 year                                (1)

S = 0.206559   R-Sq = 93.2%   R-Sq(adj) = 91.0%            (2)

Analysis of Variance                                       (3)

Source       DF    SS       MS       F       P
Regression    1   1.764   1.76400   41.34   0.008
Error         3   0.128   0.04267
Total         4   1.892
```

1. The equation is given in the form $y = a + bx$. Notice that the slope ($b$) is positive here (0.14). This tells us that the line has a positive slope; it goes from bottom left to top right.
2. MINITAB gives the standard deviation (S = square root of the residual or error mean square) and "r-squared" (R-sq). R-squared is also

commonly written $r^2$ and is called the **coefficient of determination**. These quantities make more sense after looking at the following.

3. The ANOVA table is of the same general form as for when we are comparing the effects of several treatments in an experiment (Chapter 6). However, here the "treatments" line is replaced by one called "regression." We can see that S in point 2 is the square root of the error mean square, and that the coefficient of determination is

$$r^2 = \frac{regression\_sum\_of\_squares}{total\_sum\_of\_squares}$$

Thus, $r^2$ is the proportion of the variation in $y$ accounted for by variation in $x$ (from 0 to 1, which is the same as from 0 to 100%). If it is 100%, it indicates that the regression line goes exactly through all the points on the graph, and our model then explains all the variability in the response variable: there is no random variation, so all the residuals are zero. In contrast, if $r^2$ is 0%, it is consistent with a random arrangement of points on the graph. The larger the value of $r^2$, the more useful the independent variable is likely to be as a predictor of the response variable. (Do not worry about the adjusted r-squared value on the printout. It is only helpful in more complicated models). As its name implies, $r^2$ is in fact the square of the Pearson's correlation coefficient $r$, underlining the close relationship between regression and correlation.

The constant (–272) and slope (0.14) are both estimates based on the sample of temperature measurements we have. As such, they each have a standard error, and we can calculate confidence intervals for them and test them against the Null hypothesis that they could really be zero using a t-test. To see these estimates we must use MINITAB's Regression procedure **Stat>Regression>Regression**, which gives an extra few lines of output:

```
Predictor      Coef   SE Coef      T       P
Constant    -271.96     43.22   -6.29   0.008
Year         0.14000   0.02177    6.43   0.008
```

The test statistic printed for each parameter estimate is

$$t = \frac{estimate - 0}{standard\_error} \text{ (printed in the next column).}$$

The degrees of freedom for the test are $n - 2$, where $n$ is the number of data points. The test for the constant is not particularly useful; it predicts that in the year 0 the July temperature would be –272°, which seems unlikely, and in the absence of a measurement from year 0, this conclusion can be ignored.

The test for the slope of the line is of cardinal interest, however. The t-test, $t_{3d.f.} = 6.43$, shows it to be significantly different from zero (p = 0.008). So, we can be very confident that there is a strong positive linear relationship between temperature and year.

In a simple case such as this (linear regression with only one predictor variable), the ANOVA table gives us exactly the same information as the t-test for the slope, as can be seen from the fact that the F-statistic (41.34) is exactly equal to the square of the t-value for the slope (6.43).

## 8.6    Interpreting Hypothesis Tests in Regression

In interpreting the output, it is important to realise that, just because you have shown a significant relationship, this does not necessarily mean that $x$ is *causing* the variation in $y$. Both could be responding independently to some third factor ($z$). In fact, this is probably the case here. If we think that the rise in temperature may be the result of changes in the Earth's atmosphere, then we look for some aspect of the rate of atmospheric change that accounts for the slope we observed.

In fact, at the time these data were collected, models of the Earth's climate predicted that temperatures should be increasing at 0.2°C per year. Using the information in the parameter estimates table of our analysis, we can test if our data set is consistent with this prediction using a t-test in the usual way

- *Research hypothesis:* the rate of change at our weather station differs from 0.2°C per year
- *Null hypothesis:* the rate is not different from 0.2°C per year.

$$t = \frac{0.14 - 0.2}{0.02177} = -2.76 \text{  with 3 degrees of freedom.}$$

The 5% critical value is 3.18, so we have no evidence that 0.14 differs from this value.

### 8.6.1    Fitting Data to Hypotheses

You may think that we have put the hypotheses the wrong way around; surely, the "interesting" result would be that the weather station data agrees with the predictions of the global climate model? To see why what we have done is correct, you have to think again about the logic of hypothesis testing. In Chapter 2, Section 2.4, we proposed that the Null hypothesis is *something you can disprove*. We could not prove the generalisation that the overall rate of global warming is 0.2°C per year from any number of data sets because there could always be another data set waiting to show that this was not always true. However, we can, in principle, show that 0.2°C per year is unlikely to be the true rate of warming in a particular case, by producing a data set which differs from it sufficiently to be extremely unlikely to occur if the true value actually is 0.2°C per year.

There is a general problem here with the interpretation of the "goodness of fit" of data to predictions made on the basis of some theory. Failure to reject a Null hypothesis that a calculated parameter estimate (the slope of 0.14) is close to a predicted value (0.20 in this case) could come about for two quite different reasons, namely:

- The true value actually is close to the estimate.
- We did not collect enough data for an effective test of the hypothesis.

The best advice at the moment is to use the confidence interval of the estimate as an indication of how convincing the test actually is. The standard error of the slope estimate is printed in the output (we will discuss how it was calculated in Section 8.12), so we can use this value together with the t-critical value for 3 degrees of freedom to construct a 95% confidence interval for the slope, just as we did for an estimated mean in Chapter 2 (Section 2.2).

The 95% confidence interval is

$$slope \pm t_{3d.f.} \times s.e.(slope) = 0.14 \pm 3.18 \times 0.02117 = [0.073, 0.207] \text{ to 3 s.f.}$$

You can see that the width of this interval (0.13) is quite wide relative to both the estimated value (0.14) and the hypothesis of interest (0.2). Consequently, we would not be particularly convinced by the goodness of fit to 0.2°C per year, mainly because we have very little data (only 3 d.f.). We can conclude that the true value is unlikely to be zero, at least if we stick to the rejection criterion of 0.05 (5%), because 0.14°C per year is a big enough slope to detect with only 5 data points, given the amount of variability in the data.

---

## 8.7 Further Example of Linear Regression

We used a very small sample size in the temperature example so that the basis of the calculations could be made clear, but to illustrate some further important aspects of regression analysis, we need to use a larger data set.

The amount of lead pollution on roadside verges is likely to depend on the amount of traffic passing along the road. In a survey, the amount of lead in parts per million ($y$) on the vegetation alongside 19 roads was recorded, together with the amount of traffic in hundreds of vehicles passing per day ($x$).

We can enter the data into two columns of a MINITAB worksheet:

| ROW | Lead | Traffic |
|-----|------|---------|
| 1 | 44 | 79 |
| 2 | 58 | 109 |
| 3 | 43 | 90 |
| 4 | 60 | 104 |
| 5 | 23 | 57 |
| 6 | 53 | 111 |
| 7 | 48 | 124 |
| 8 | 74 | 127 |
| 9 | 14 | 54 |
| 10 | 38 | 102 |
| 11 | 50 | 121 |
| 12 | 55 | 118 |
| 13 | 14 | 35 |
| 14 | 67 | 131 |
| 15 | 66 | 135 |
| 16 | 18 | 70 |
| 17 | 32 | 90 |
| 18 | 20 | 50 |
| 19 | 30 | 70 |

We first make a scatter plot, with lead (the response variable) on the y axis (Figure 8.6). There appears to be a positive relationship. We now run a regression analysis. In MINITAB we use **Stat>Regression>Regression** as we

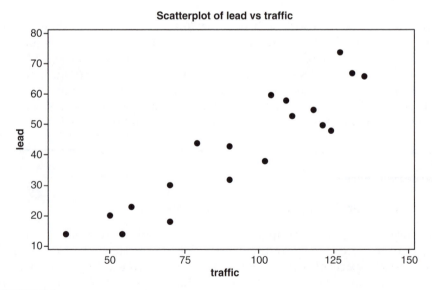

**FIGURE 8.6**
Scatter plot of lead against traffic.

want to see the coefficients and their standard errors. Looking through the output, we find:

```
The regression equation is
lead = -10.7 + 0.569 traffic
```

- What is the intercept? What is the slope? (10.7, 0.569)

```
Predictor        Coef   Stdev     t-ratio       p
Constant      -10.736   5.813       -1.85   0.082
Traffic       0.56893  0.05923       9.61   0.000
```

- Is the intercept significantly different from zero? (No, $t_{17 d.f.} = -1.85$, $p = 0.082$.) (Note that the estimate of $-10.736$ for the intercept must be meaningless as we cannot have a negative value for lead. It may well be that if our sample included lower levels of traffic, we would find that the relationship was curved, Figure 8.7.)
- Is the slope significantly different from zero? (Yes, $t_{17 d.f.} = 9.61$, $p < 0.0005$.)

```
s = 7.680    R-sq = 84.4%    R-sq(adj) = 83.5%
```

- Is this equation of a straight line a useful model to describe the relationship between roadside lead and traffic numbers? (Yes, it accounts for 84.4% of the variance in lead levels.)

```
Analysis of Variance:
SOURCE        DF     SS      MS       F       p
Regression     1   5442.0  5442.0   92.26   0.000
Error         17   1002.8    59.0
Total         18   6444.7
```

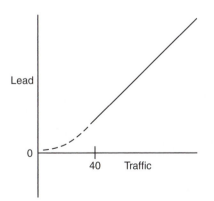

**FIGURE 8.7**
Relationship between lead and traffic.

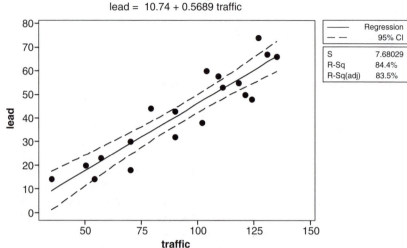

**FIGURE 8.8**

Fitted line plot of lead against traffic.

- With what confidence can you reject the Null hypothesis of no linear relationship between lead concentration and amount of traffic? ($F_{1,17}$ = 92.96, p < 0.0005), so we would reject the Null hypothesis of no effect even if we choose a rejection criterion of 0.001 or 0.1%. This appears to be very strong evidence against the Null hypothesis of no effect.)

A plot of the observations with the regression line superimposed is shown in Figure 8.8, which also shows the confidence interval for the fitted line (see Section 8.6.1 and Section 8.10).

## 8.8 Assumptions

The assumptions required for correct conclusions to be drawn from a regression analysis are essentially the same as those required for ANOVA (Chapter 6, Section 6.10), but they manifest themselves in slightly different ways because the predictor variable is quantitative.

### 8.8.1 Independence

Every observation of $y$ (lead concentration in the MINITAB example) must be independent of every other $y$. If, for example, we included several measurements from the same road, they would not represent independent tests of the hypothesis that the traffic density on roads in general has an effect on lead pollution.

We need to ensure independence by carrying out our sampling or experimentation in an appropriate way.

### 8.8.2 Normal Distribution of Error

A Normal distribution of residuals (errors) is assumed, as they represent random variation. There should be many residuals with very small absolute values (near zero), and only a few with very large ones (far from zero). We can test this by asking for residual plots in MINITAB. When using **either** **Stat>Regression>Regression** or **Stat>Regression> Fitted Line**, select the Graph button and chose Histogram of Residuals. In the lead/traffic example, we find that the residuals have an approximately Normal distribution albeit with a slight kink in the 5 class. (Figure 8.9a). The Normality plot (Figure 8.9b) shows a sufficiently straight line for us to accept that the residuals have a Normal distribution (cf. Chapter 6, Section 6.10.3).

### 8.8.3 Homogeneity of Variance

Homogeneity of variance of residuals is assumed. The residuals should not show a tendency to increase (or decrease) as $x$ increases. We can examine this by plotting residuals against the fitted values; in MINITAB we chose the Graph Option **Residuals vs. Fits** (Figure 8.10). This is fine; there is no obvious pattern. A common problem is shown in Figure 8.11, where there is much more variability at high levels of $x$ than at low levels. If this occurs, consult a statistician. It may be possible to correct this situation by changing the scale of one or both the axes (statistical **transformation,** see Grafen and Hails, 2002).

### 8.8.4 Linearity

In linear regression, it must be plausible that the relationship between the response and predictor is a straight line. If the true relationship is a curve, the model is not a meaningful description of the data, and conclusions about the slope of a linear relationship are of little value. Looking at the scatter plot with the fitted line on it (Figure 8.8) gives a good indication whether the linear model is sensible. This point is discussed further in Section 8.10 (Figure 8.12b).

### 8.8.5 Continuity

Ideally, both $x$- and $y$-variables should be measured on a continuous scale (like kilograms, minutes, or centimetres) rather than being counts, proportions, or

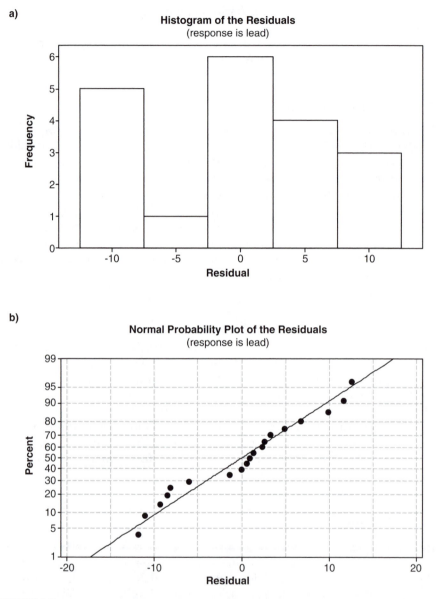

**FIGURE 8.9**
Residuals plots for lead against traffic: (a) histogram of residuals. Reasonably normal (b) Normality plot. No systematic departure from Normality.

ranks. Counts and proportions can also often be analysed using regression but are likely to violate the homogeneity of variance assumption (see Section 8.8.3). It is important to examine the graph of the residuals and, if in any doubt, you should ask advice. Other special techniques for ranked observations are introduced in Chapter 10.

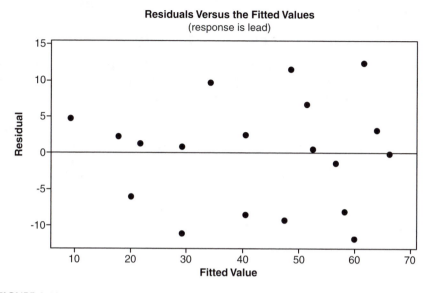

**FIGURE 8.10**
Residuals against fitted values for lead against traffic.

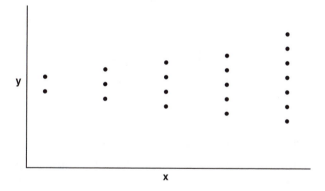

**FIGURE 8.11**
Variability in $y$ increases with variability in $x$: the "shotgun effect."

### 8.8.6 Absence of "Errors" in x-Values

Ideally, the $x$-observations should be "without error." Obviously, we want our measurement of the $x$-values to be as accurate as possible, although we realise that they will not be perfect because of measurement errors. If there is no bias, slight mistakes either way in measuring $x$-values will merely increase the residual mean square and make the regression less precise. However, if there is reason to suspect bias in the observation of $x$, such that it is generally recorded as too high (or too low), this will affect the estimation of the slope.

## 8.9    The Importance of Plotting Observations

The lead deposition/traffic data set was a typical example of a relationship that is well modeled by linear regression. However, it is possible for the computer to give you the same equation (and the same values for $p$ and $r^2$) from very different data sets, for some of which the model would be inappropriate. This is an alarming thought. It is important to make plots of the observations, and of the residuals plotted against the fitted values, to identify such problems.

We will use four data sets devised by Anscombe (1973) to illustrate this problem. The $x$-values for data sets 1, 2, and 3 are the same and are in column "x123" whereas that for data set 4 is in column "x4." The corresponding responses are in columns named y1, y2, y3, and y4:

| Row | x123 | y1 | y2 | y3 | x4 | y4 |
|-----|------|-------|------|-------|----|------|
| 1 | 10 | 8.04 | 9.14 | 7.46 | 8 | 6.58 |
| 2 | 8 | 6.95 | 8.14 | 6.77 | 8 | 5.76 |
| 3 | 13 | 7.58 | 8.74 | 12.70 | 8 | 7.71 |
| 4 | 9 | 8.81 | 8.77 | 7.11 | 8 | 8.84 |
| 5 | 11 | 8.33 | 9.26 | 7.81 | 8 | 8.47 |
| 6 | 14 | 9.96 | 8.10 | 8.84 | 8 | 7.04 |
| 7 | 6 | 7.24 | 6.13 | 6.08 | 8 | 5.25 |
| 8 | 4 | 4.26 | 3.10 | 5.39 | 19 | 2.50 |
| 9 | 12 | 10.84 | 9.13 | 8.15 | 8 | 5.56 |
| 10 | 7 | 4.82 | 7.26 | 6.42 | 8 | 7.91 |
| 11 | 5 | 5.68 | 4.74 | 5.73 | 8 | 6.89 |

### 8.9.1    Same Equation from Different Patterns?

In Figure 8.12, we show the scatter plots with the best fitting regression lines for the four data sets. The patterns made by the data points are obviously very different. If we give them names, we might call them:

a) A straight-line relationship with some scatter.

b) A curve.

c) A straight line with one point having a very different y-value from the others.

d) A straight line with one point having a very different x-value from the others.

The first point to notice is that the best-fit equation, $Y_n = 3.00 + 0.5*X_n$, is the same (give or take small rounding errors) for all four data sets. More surprisingly, perhaps, the values of S and R-squared are also the same (Figure 8.12); as we have seen, we can deduce all the other figures generated by the

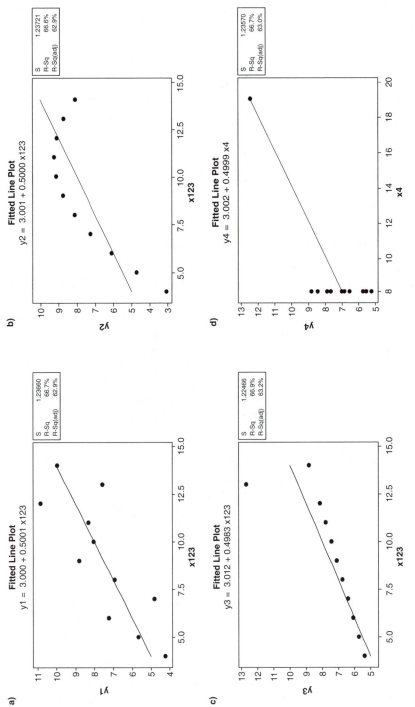

**FIGURE 8.12**
Fitted lines for (a) data set 1, (b) data set 2, (c) data set 3, and (d) data set 4.

output from these, so we omit the full printout for brevity. On the basis of the calculations alone we might not realise that there was a problem. What should we do?

### 8.9.2 Plot the Data and the Fitted Line

We have already done this in Figure 8.12, and inspection of how well the line appears to describe the data is often a helpful guide as to whether the model is correct. Clearly, in data set 2, the straight line is not appropriate to the curved data; here, the assumption of linearity is violated. In data set 3 and data set 4, the outlying points obviously have a very important effect on the position of the line. If we eliminated them, then the estimates of the slope and intercept would be very different.

### 8.9.3 Plot the Histogram of the Standardised Residuals

These are shown in Figure 8.13. Recall that it is an assumption that the residuals (errors) are Normally distributed. If we use standardised residuals, they are in units of standard error, so if the distribution is Normal, its standard deviation is 1 (see Chapter 1, Section 1.5.1, if you have forgotten about this). The graphs allow us to check informally that (1) the distribution is symmetrical about zero, and (2) the majority of the residuals (about 95% of them) lie in the range +/−1.96). This appears to be the case in data set 1 and data set 4, but there are clearly problems with symmetry in data set 2 and data set 3. This would indicate that the straight-line model was unsatisfactory for data set 2 and data set 3, though by itself it would not tell us precisely where the problem lay. We might not realise there was a problem with dataset 4 here.

A similar story can be seen with the Normality plots of the standardised residuals (Figure 8.14). The plots for data set 1 and data set 4 are satisfactorily straight, but in dataset 2 the points are to the left of the line at the ends and to the right in the middle, which would not be satisfactory, whereas in data set 3 the outlier clearly represents a departure from Normality.

### 8.9.4 Plot the Residuals against the Fitted Values

These are shown in Figure 8.15. Formally, this is a test of the homogeneity of the variances; what it means in practical terms is that deviations of the data from the fitted model should be approximately the same for any fitted value. This appears to be the case in data set 1 but is clearly not in any of the others. Residual patterns like Figure 8.15b, where all the residuals at the ends of the x-axis are negative, and all the ones in the middle are positive, usually indicate a **structural error** in the model. Here, we should have fitted a curve rather than a straight line.

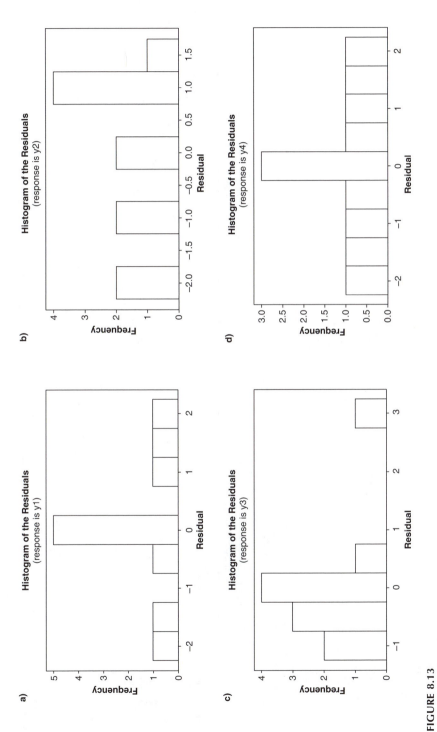

**FIGURE 8.13**
Histograms of residuals for (a) data set 1, (b) data set 2, (c) data set 3, and (d) data set 4.

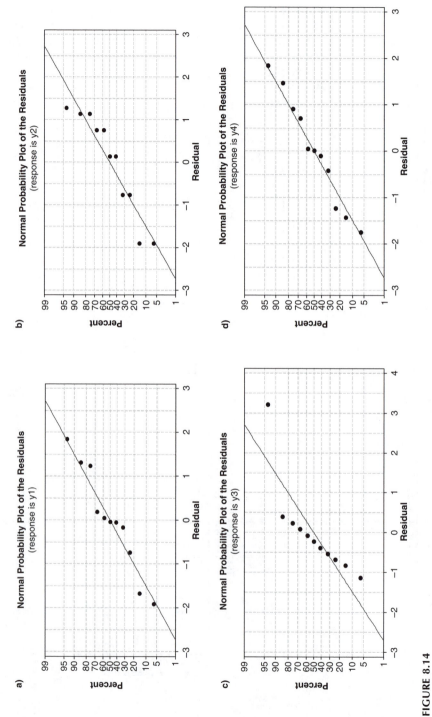

**FIGURE 8.14**
Normality probability plots of the residuals for (a) data set 1, (b) data set 2, (c) data set 3, and (d) data set 4.

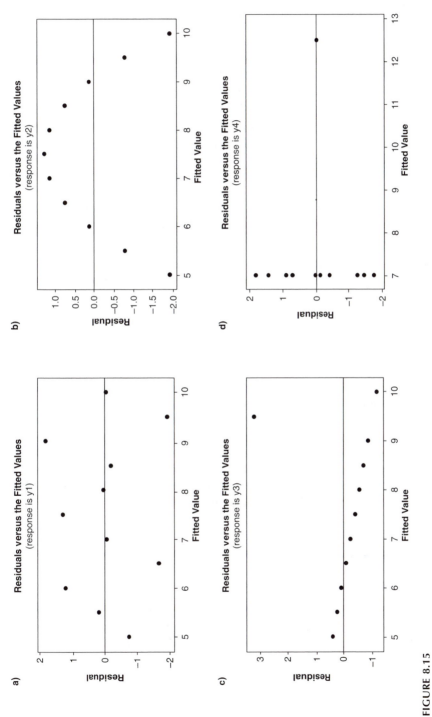

**FIGURE 8.15**
Residuals against fitted values for (a) data set 1, (b) data set 2, (c) data set 3, and (d) data set 4.

most probably lies, we have to take into account the uncertainty about both the slope and the intercept. The confidence interval for each fitted value is usually given by the rather daunting-looking formula

$$y' \pm t \times \sqrt{EMS \times \left( \frac{1}{n} + \frac{(x' - \bar{x})^2}{SSX} \right)}$$

where $y$ is the fitted value and $x$ is the value of $x$ for which the fit has been calculated.

However, it makes better sense to express it in a slightly different but equivalent way (multiplying out the bracketed term inside the square root):

$$y' \pm t \times \sqrt{\left( \frac{EMS}{n} + \frac{EMS \times (x' - \bar{x})^2}{SSX} \right)}$$

Here, we can see that the standard error part (the square-root term) is the sum of two components of variation. The first bit is the variance in the intercept and is analogous to $\sigma^2/n$ for the sampling variance of a mean (Chapter 2). The second bit multiplies the EMS by the difference between the distance that the $x$-value involved in each prediction lies from the mean of the $x$-values. This makes sense because we know that the best-fit line *must* pass through the point $\bar{x}, \bar{y}$, so the contribution of the slope term to the variation in the predicted $y$ is zero when predicting the $y$-value for $\bar{x}$. When we ask for the confidence interval to be plotted, we are seeing each fitted value (on the line) with the confidence interval around it (this is seen in Figure 8.8). We are 95% confident that the line that represents the relationship for the entire population is within this range. Notice how the confidence interval is narrowest in the centre of the graph and becomes wider at the extremes. This follows from the dependence of the confidence interval on $(x' - \bar{x})^2$, but intuitively, it is because we know less about the relationship at the extremes than we do in the middle. So, as we approach this area, we become less and less certain about the location of the line.

## 8.11  Standard Error of Prediction (Prediction Interval)

What would be the likely values we might expect if further samples were taken from the same population? This is the prediction interval. We now have to take into account the fact that further data would be subject to sampling error as well as to the uncertainty about the values of the parameters. If you take a simple example of picking a new value from a population whose parameters we have estimated from a sample, even if you knew the true mean and standard deviation of the population, there would still be uncertainty about the value you picked. In fact, you could think of it as another sample of size 1, so the sampling distribution for it would have standard deviation

$$\sigma/\sqrt{n} = \sigma/1 = \sigma$$

in the usual way for a sampling distribution. As we saw already, the $\sqrt{EMS}$ plays the role of the population standard deviation, so we just add $\sqrt{EMS}$ as an additional term to get

$$y' \pm t \times \sqrt{\left( EMS + \frac{EMS}{n} + \frac{EMS \times (x' - \bar{x})^2}{SSX} \right)}$$

Or, more conventionally (factorising),

$$y' \pm t \times \sqrt{EMS \times \left( 1 + \frac{1}{n} + \frac{(x' - \bar{x})^2}{SSX} \right)}$$

This is the interval we see if we ask MINITAB for the prediction interval. It is obviously wider than the confidence interval for the line. Figure 8.16 shows both intervals plotted for the lead/traffic data set.

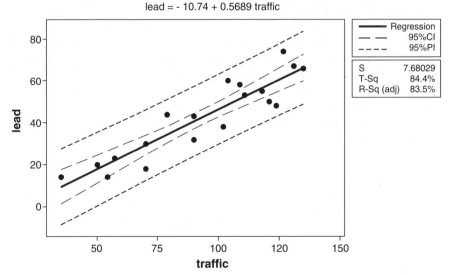

**FIGURE 8.16**
Fitted line for lead vs. traffic showing Prediction Interval as well as Confidence Interval for the line.

Again, this follows the usual format: estimate plus or minus t times the standard error of that estimate. It is just that here the standard error is more complex.

---

## 8.12 Caution in the Interpretation of Regression and Correlation

### 8.12.1 Correlation Is Not Causation

It is commonly found that there is a strong positive linear relation between the number of doctors in a city and the number of deaths per year in that city! At first sight we may be tempted to conclude that having more doctors leads to more deaths. Therefore, if we cut the number of doctors, we might expect fewer deaths! However, we have overlooked the fact that both the number of doctors and the number of deaths in a city depend upon the population of a city. We can calculate correlations between any pair of variables, but we must always be wary of assuming that one causes variation in the other.

### 8.12.2 Extrapolation Is Dangerous!

If we predict a y-value for an x-value outside our range of data (**extrapolation**), we cannot be sure that the relationship will still be of the same form. For example, a plant that is growing steadily faster over the range of fertiliser from 0 to 300 kg/ha is unlikely to continue growing at the same rate if fertiliser were applied at the rate of 1000 kg/ha. It will probably show signs of toxicity, but we actually have no evidence about its behaviour at this dosage.

A similar point can be made about the Alaska temperature example (Section 8.4). Just because the rate of change of July temperatures was 0.14°C/year between 1979 and 1991, this does not mean it will continue to rise at the same rate for the next 12 years. It may do so, but we have no evidence about it. Extrapolation into the future in this way is surprisingly common (think of predictions about global population growth) but should be made with extreme caution, if at all.

# 9

## Categorical Data

> Not everything that can be counted counts; and not everything that counts can be counted.

> —**George Gallup**

Until now, we have shown you how to analyse data where the response variable is *quantitative*. These are continuous data, such as weights or lengths, which can have intermediate values. It also includes *count* data (the number of individuals or events). Such measurements have both magnitude and order. Three is more than two, which is more than one. This is not true of categorical data. Categorical data are data that fit into one of several categories such as blue, green, or red, or vertebrate or invertebrate. Red is not more red than blue; it is just different.

We already know about categorical variables because, in Chapter 3 to Chapter 7, the predictor variables were categories (for example, old diet, new diet) even though the questions we were asking concerned quantitative response variables such as weight. In this chapter we look at the case where both the response and the predictor variables are categorical. This may seem an odd way to describe what we are about to do, which comes in two forms:

1. If we count objects up and classify them as belonging to a category, do they fall in the proportions predicted by an hypothesis? For example, if we have 12 children classified by gender, do they occur in the proportion predicted by the hypothesis of an equal sex ratio? These are **goodness-of-fit tests** (Section 9.1).

2. If we can classify objects in two different ways, do the proportions according to one classification fall independently of those according to the other? For example, does the proportion of students studying physics who are female differ from the proportion studying biology who are female? These are **contingency tests** (Section 9.3).

Although methods for categorical variables are often seen as a "special case" without much connection to the rest of the methods we have learned in this book, there is, in fact, a common thread, and we discuss briefly how they fit into the general pattern, in Section 9.5.

## 9.1   The Chi-Squared Goodness-of-Fit Test

### 9.1.1   Simple Example: Sex Ratio in an Oxford Family

In November 2000, the *Daily Mail* reported the birth of a male baby to an Oxford mother who already had 11 daughters. Does a sex ratio as extreme as 11:1 girls:boys represent a departure from the expected sex ratio of 50:50? We might expect that a family of 12 should consist of 6 girls and 6 boys, but there would be some variation about this. There is a discussion of the theory behind these expectations in Chapter 1. Here the question is whether the discrepancy between the expectation of 6 girls and 6 boys and the observation of 11 girls and one boy is large enough to lead us to doubt the assumption that a boy or girl is equally likely for this particular mother.

The idea behind this analysis is that we look at the discrepancies between the observed and expected frequencies. For each difference we calculate the quantity

$$\frac{(observed - expected)^2}{expected}$$

In view of the discussion of sums-of-square in Chapter 6, it may be no surprise that we square the differences between observed and expected frequencies; dividing the difference by the expected is less easy to explain, so suffice it to say that this move generates a standardised sum-of-squares that can be used as a test statistic. The calculations we need are presented in Table 9.1. Like other sums-of-squares, the last column can be added together to give a quantity that has a known distribution if the Null hypothesis is true. This is the test statistic for the problem; the value is 8.3333. The test statistic follows a distribution known as chi-squared (also written $\chi^2$). As with tests we are familiar with from earlier chapters, we reject the Null hypothesis if the test statistic exceeds a critical value.

Formally our procedure is

- Research hypothesis: The frequencies of boys and girls depart from the expected frequencies when the sex ratio is 50:50.
- Null hypothesis: No reason to believe 50:50 sex ratio does not apply; departures are just due to sampling error.
- Test statistic: $\chi^2$ as calculated in Table 9.1.
- Find probability of test statistic if $H_0$ is true. Look up in Table C.6 (or use computer).

As with the F and t statistics, we have to establish the correct number of degrees of freedom for the test. It is the *number of categories* −1; here, it is

**TABLE 9.1**

Counts of Observed Gender in a Family of 12 Children on the Hypothesis that Boys and Girls Are Equally Likely

| Gender | Observed | Expected | Chi-Square |
|--------|----------|----------|------------|
| Boy | 1 | 6 | 4.16667 |
| Girl | 11 | 6 | 4.16667 |

$2 - 1 = 1$ d.f. If we look in Appendix C, Table C.6, we find that, for 1 d.f. and a rejection probability of 5%, $H_0$ should be rejected if the test statistic exceeds 3.84. As 8.3333 is much larger than 3.84, we would reject $H_0$ here. A family of 11 girls and one boy is extremely unlikely to occur if the underlying probability of a girl and a boy live birth are equal, so we would suspect that there is some factor in this family leading to a bias in the sex ratio.

The latest release (15) of MINITAB provides this test as **Stat>Tables> Chi_Squared Goodness-of-Fit test.**

```
Chi-Square Goodness-of-Fit Test for Observed Counts in Variable: n

Using category names in gender

                      Historical        Test              Contribution
Category  Observed      Counts   Proportion  Expected        to Chi-Sq
       f        11         0.5          0.5         6          4.16667
       m         1         0.5          0.5         6          4.16667

 N   DF   Chi-Sq   p-Value
12    1  8.33333     0.004
```

The printout of the program shows the same calculations, with the exact probability for a chi-squared value of 8.3333 with 1 d.f., p = 0.004. In earlier releases, the calculations must be done by hand (example in the next section).

## 9.2 A More Interesting Example: Testing Genetic Models

In analysing genetic data, we frequently find we must estimate the gene frequencies from the data before we can test hypotheses about the frequencies with which different genotypes would be expected to occur under a Null hypothesis.

It has been known for a long time that sickle cell anemia, a mutation in the gene controlling one of the proteins that make up haemoglobin in humans, is more common in areas where malaria is endemic. The sickle cell trait is recessive but, although it is not fully understood why, heterozygous individuals are known to have some protection against the symptoms of malaria. This advantage is sufficient to maintain the sickle cell alleles in such populations, even though the homozygotes suffer from the debilitating disease sickle cell anemia. Because the heterozygotes can be recognized from a blood test, it is possible to establish at what frequency sickle cell alleles exist in a population.

A recently collected sample of blood tests from 886 randomly chosen individuals in Gabon contained the following genotypes:

- Normal/normal (A/A) 700
- Normal/sickle (A/S) 180
- Sickle/sickle (S/S) 6

The gene for the normal protein is called haemoglobin A, and that for sickle cell is called haemoglobin S. Because an A/S individual contains exactly one S allele, and an S/S individual contains 2, we can say that there are $180 \times 1 + 6 \times 2 = 192$ S alleles out of a total of $886 \times 2 = 1772$ alleles in this sample. Thus, our estimate of the frequency of the S allele is $192/1772 = 0.108$ of the gene pool, which means that the frequency of the A allele is $1 - 0.108 = 0.892$.

### 9.2.1  Expected Proportions

An elegant but simple bit of mathematical theory allows us to predict the expected proportions of A/A, A/S, and S/S genotypes based on the proportions of S and A alleles in the population. It is best illustrated in a table of each of the possible combinations (Table 9.2):

These figures allow us to estimate what the phenotype frequencies would be after one further generation provided the following were true:

- Random mating between individuals of all genotypes.
- No selection is operating.

What is not quite so obvious is that these phenotypic frequencies would be *attained after only a single generation*, provided the assumptions about mating and selection were correct. Provided matings continued to be random and selection weak or absent, they would then remain at these frequencies

**TABLE 9.2**

Phenotype frequencies Calculated for Sickle Cell Trait from the Estimated Gene Frequencies

|                | A (p = 0.892) | S (q = 0.108) |
|----------------|---------------|---------------|
| A (p = 0.892)  | A/A $(p^2 = 0.892^2 = 0.796)$ | A/S $(p \times q = 0.892 \times 0.108 = .096)$ |
| S (q = 0.108)  | S/A $(q \times p = 0.108 \times 0.892 = 0.096)$ | S/S $(q^2 = 0.108^2 = 0.012)$ |

(see Box 9.1). This is known as the **Hardy–Weinberg equilibrium** and allows us to say that a population will deviate from the expected frequencies shown in Table 9.2 only if there is either assortative mating or selection (or both). The requirement of random mating can often be checked independently by looking to see who actually mates with whom, so this provides an important method for detecting natural selection using genetic data from the field (Hedrick, 2005).

---

## BOX 9.1    Why the Hardy–Weinberg Ratio Represents an Equilibrium

Consider a population of arbitrary size, with the unrealistic but simple Hemoglobin-relative genotype frequencies for sickle cell anemia:

| A/A | A/S | S/S |
|-----|-----|-----|
| 0.4 | 0.4 | 0.2 |

Work out the allele frequencies as usual, writing p for the frequency of A and q for the frequency of S:

$$p = \frac{0.4 \times 2 + 0.4}{2} = 0.6$$

$$q = \frac{0.4 + 0.2 \times 2}{2} = 0.4$$

Assuming random mating and no selection, the next generation will have the following genotypes:

| A/A | A/S | S/S |
|-----|-----|-----|
| $p^2 = 0.6^2 = 0.36$ | $2pq = 2 \times 0.6 \times 0.4 = 0.48$ | $q^2 = 0.4^2 = 0.16$ |

Gene frequencies will now be

$$p = \frac{0.36 \times 2 + 0.48}{2} = 0.6$$

$$q = \frac{0.48 + 0.16 \times 2}{2} = 0.4$$

which is the same as before. Thus, all subsequent generations will have p = p(A) = 0.36 and q = p(B) = 0.4, so the phenotypes will remain in the ration 0.36:0.48:0.16 as long as there is no selection and no assortative mating.

---

### 9.2.2 Expected Frequencies

Having estimated the proportion of the S allele as 0.108, we can now proceed to test the Null hypothesis that the population is in Hardy–Weinberg equilibrium. Given that it is known that matings do not depend on sickle cell phenotype, rejecting the Null hypothesis may allow us to see selection in action. Here we do the calculations by hand (Table 9.3).

In the table, we calculate the expected proportions from the Hardy–Weinberg theorem and then calculate the expected frequencies by multiplying by the total number of individuals. These are compared to the observed frequencies as before using the calculation

$$\frac{(observed - expected)^2}{expected}$$

for each row. The total of the last column is the calculated $\chi^2$ for the test.

### 9.2.3 Degrees of Freedom — A Catch

You might expect that the degrees of freedom for this test would be $n - 1 = 2$ for the $n = 3$ classes. However, we must subtract an additional degree of freedom because we estimated the allele frequency from the data. Thus, here, we look up the critical value of $\chi^2$ for *one degree of freedom* $(n - 1 - 1)$. The calculated value 2.631 is somewhat less than the critical value for 1 d.f. of 3.84, so we can not reject the Null hypothesis that these individuals are in Hardy–Weinberg equilibrium. Thus, there is, in fact, no evidence for selection either for or against the S allele in this population.

At the time of writing the MINITAB Release 15, goodness-of-fit procedure does not allow us to specify the loss of an additional degree of freedom to take account of the estimation of a parameter (in this case the allele frequency) from the data.

**TABLE 9.3**

Goodness-of-Fit Test to Hardy–Weinberg Equilibrium for the Phenotype Frequencies Observed in Gabon, Using Gene Frequencies Estimated from the Data

| Phenotype | Proportion (Calculated in Table 9.2) | Expected Frequency | Observed Frequency | Chi-Squared |
|---|---|---|---|---|
| A/A | $p^2 = 0.796$ | $0.796 \times 886 = 705.26$ | 700 | 0.039 |
| A/S | $2p \times q = 0.192$ | $0.192 \times 886 = 170.11$ | 180 | 0.575 |
| S/S | $q^2 = 0.012$ | $0.012 \times 886 = 10.63$ | 6 | 2.017 |
| Total | 1.000 | 886 | 886 | **2.631** |

### 9.2.4 Hypothesis Testing in Genetics

Formally our procedure is:

- Research hypothesis: the frequencies of different offspring types departs from the predictions of the Hardy–Weinberg model.
- Null hypothesis: no reason to believe the basic genetic model does not apply. Frequencies will be in accordance with the model.
- Test statistic: $\chi^2$ as calculated in the table.
- Look up in tables (or use computer to calculate) probability of the test statistic value if $H_0$ is true).

You may think this is a strange way around: the Null hypothesis is that the population is described by the Hardy–Weinberg model, whereas you might imagine that we want to test the Hardy–Weinberg model to see if it works with real data. Recall that in Chapter 8, Section 8.6, we discussed the problems of confirming theories from goodness-of-fit tests; you can only reject a theory on the basis of a goodness-of-fit test, never confirm it. Thus, especially in analysing genetics data, we tend to find that we are looking for ways in which the data *depart from* the theoretical expectations based on a model of how the genetic process is operating. Such departures are the interesting results, as they show us where the "standard" model breaks down, and consequently, where something interesting is happening.

## 9.3 Contingency Analysis: Chi-Squared Test of Proportions

The other common use of the chi-squared distribution is called **contingency analysis**. Consider the case where we can classify subjects according to two different characteristics. The classifications represent the effects of different characteristics on subjects. What we want to test is whether the proportion of subjects in one classification depends on which class they are in the other. Let us clarify this with an example:

> A new antimalarial drug has been developed that may be more or less effective at clearing all parasites from the blood of humans within 36 h. In an experiment, 287 individuals took part to compare its effectiveness with that of chloroquine (the standard). Of the 184 individuals receiving chloroquine, 129 were cleared of parasites within 36 h, whereas 55 were not. We can summarise these observations and the corresponding figures for the new drug in a table of *observed* values (O) (Table 9.4).

**TABLE 9.4**

Number of Patients Whose Blood Was Cleared, or Not Cleared,
of Malaria Parasites, Compared for Chloroquine and a New Drug

|             | Cleared in 36 h | Not Cleared in 36 h | Total |
|-------------|-----------------|---------------------|-------|
| Chloroquine | 129             | 55                  | 184   |
| New drug    | 80              | 23                  | 103   |
| Total       | 209             | 78                  | 287   |

Note that the numbers of individuals taking the new drug and those taking the old one do not have to be equal, although the analysis will be more robust if they are of similar magnitude. Our Null hypothesis is that the two variables are statistically independent: the proportion of individuals from whose blood the parasite has been cleared is the same for both drugs. The first step is to calculate the number of individuals we would expect to be in each category (or "class" or "cell") if the Null hypothesis is true. The maximum number of people who could appear in the top left cell is 184 because that is the number who were treated with chloroquine. The proportion 209/287 is the overall proportion of individuals from whose blood the parasite had cleared in 36 h, irrespective of which drug they had received. This proportion is then multiplied by the number of individuals who received chloroquine (184) to give the number of individuals we would *expect* to find who had both received chloroquine and whose blood had cleared in 36 h, assuming both drugs have the same effect. This value is what we would *expect* if the Null hypothesis is true.

For the top left cell of Table 9.4 (chloroquine, cleared), this is

$$Expected\ value = E = \frac{209}{287} \times 184 = 133.99$$

In general, the expected number in each cell is

$$Expected\ value = \frac{column\ total}{grand\ total} \times rowtotal$$

The test is called a chi-squared **contingency** test because the expected values for one classification are *contingent* upon (dependent upon) the classification of the other.

The table of such expected values (E) is then (Table 9.5) as follows. The general idea is that if the Null hypothesis is true, the observed and expected counts will be very similar, whereas if it is false, they will be very different.

**TABLE 9.5**

Expected Number of Patients Whose Blood Would Be Cleared or Not Cleared, Calculated from the Observations in Table 9.4

|  | Cleared in 36 h | Not Cleared in 36 h | Total |
|---|---|---|---|
| Chloroquine | 133.99 | 50.01 | 184 |
| New drug | 75.01 | 27.99 | 103 |
| Total | 209 | 78 | 287 |

To enable us to calculate the strength of evidence against the Null hypothesis, we calculate the chi-squared test statistic from the observed (O) and expected (E) frequencies (similarly to the goodness-of-fit test):

$$\chi^2 = \Sigma \frac{(O - E)^2}{E}$$

Here we have

$$\chi^2 = (129 - 133.99)^2 / (133.99) + (55 - 50.01)^2 / (50.01)$$

$$+ (80 - 75.01)^2 / (75.01) + (23 - 27.99)^2 / (27.99) = 1.91$$

As always, we need to know the degrees of freedom to find the critical value in statistical tables. In a contingency table there are $(r - 1) \times (c - 1)$ degrees of freedom where $r$ = the number of rows and $c$ = the number of columns. Thus, for a $5 \times 4$ table, the d.f. = 12. Here, in our $2 \times 2$, table the d.f. = 1. We compare the calculated $\chi^2$ (1.91) with a critical value from Table C.6 taken from the column headed 95% (p = 0.05) and the row for 1 d.f. (3.841). As 1.91 is less than the table value of 3.841, we cannot reject the Null hypothesis of no difference in the efficacy of the two drugs.

### 9.3.1 Organizing the Data for Contingency Analysis

MINITAB (in common with most other packages) allows frequency data to be arranged in different ways for a chi-squared test. The "quick and dirty" method is just to put the frequency of each class in a separate cell in the worksheet thus:

```
MTB > print c11 c12

Row   Clear   Not
 1     129     55
 2      80     23
```

We then use **Stat>Tables>Chi-Square Test (Table in Worksheet)**

```
Chi-Square Test: clear, not

Expected counts are printed below observed counts
Chi-square contributions are printed below expected counts

          Clear    Not  Total
1           129     55    184
         133.99  50.01
          0.186  0.499
2            80     23    103
          75.01  27.99
          0.332  0.891

Total       209     78    287


Chi-sq = 1.908, d.f. = 1, p-value = 0.167
```

The chi-square value is the same as we calculated by hand, and MINITAB provides an exact probability that, because it is greater than our 5% criterion, leads us not to reject the Null hypothesis and conclude that there is no evidence that the new drug is an improvement on chloroquine.

A more flexible alternative is to set up the data as an "edge table" (see Appendix A, Section A.7):

```
Data Display

Row  Patients   Result          Drug
 1        129   Cleared  Chloroquine
 2         80   Cleared          New
 3         55       Not  Chloroquine
 4         23       Not          New
```

Here we have a column for the frequency of each group, a column saying which result was obtained, and a column saying which drug was used. We then use the command **Stat>Tables>Cross Tabulation and Chisquare For rows:** drug **For columns** result **Frequencies ...** patients **Chi-Square ... Display: Chi-Squared Analysis, Expected Cell Count.** This is what we get:

```
Tabulated statistics: drug, result

Using frequencies in patients

Rows: drug    Columns: result

              Cleared     Not     All
Chloroquine       129      55     184
                134.0    50.0   184.0
               0.1861  0.4985       *
New                80      23     103
                 75.0    28.0   103.0
               0.3324  0.8906       *

All               209      78     287
                209.0    78.0   287.0
                    *       *       *

Cell contents: count
               Expected count
               Contribution to chi-square

Pearson chi-square = 1.908, d.f. = 1, p-value = 0.167
Likelihood ratio chi-square = 1.945, d.f. = 1, p-value = 0.163
```

MINITAB prints the expected values in each category below the observed values, and presents the row and column totals and grand total. The third number in each cell is the contribution of that cell to the overall $\chi^2$ (that is, the value of "observed — expected," squared and divided by expected for that cell), which are added up to give the chi-squared statistic; in the edge table version, we must remember to ask for these. The overall $\chi^2$ is printed together with its degrees of freedom at the bottom of the table. Here, the $\chi^2$ statistic is very small, indicating that the observed values are very close to those we expected to find if the Null hypothesis were true. As 1.908 is less than the critical value of $\chi^2_{1df}$ (3.84), we conclude that a similar proportion of people are cured by the two drugs. MINITAB gives us the exact value of p = 0.167.

### 9.3.2 Some Important Requirements for Carrying Out a Valid $\chi^2$ Contingency Test

- The observations must be counts *of independent events or individuals*. They *must not* be percentages, proportions, or measurements.
- In general, the *expected* count or frequency in each class should exceed two, and also, 80% of the classes should have expected frequencies greater than five. If this is not so, then either we should collect more data, or we can combine neighbouring classes (if this is sensible).

For example, if we applied fertiliser to 30 of 60 plants at random and classified the plants after one month's growth, we might find the following (Table 9.6a).

**TABLE 9.6a**

Growth Category of 60 Plants Classified by Whether
They Received a Fertiliser or Not

|                | Small | Medium | Large |
|----------------|-------|--------|-------|
| No fertiliser  | 20    | 7      | 3     |
| With fertiliser| 10    | 15     | 5     |

*Note:* With the data in this form the expected values for the
large category will be less than 5.

**TABLE 9.6b**

Rearrangement of the Data in Table 9.6a

|                 | Small | Large |
|-----------------|-------|-------|
| No fertiliser   | 20    | 10    |
| With fertiliser | 10    | 20    |

*Note:* By combining the medium and large categories into
a single category, "Large," we have a table in which
all expected values are > 5.

The expected value (Section 9.3) for the top right class (large, no fertiliser)
would be

$$\frac{column\ total}{grand\ total} \times row\ total = \frac{8}{60} \times 30 = 4$$

For the bottom right class (large, with fertiliser) the expected value would
be the same. To fulfill the requirements for a $\chi^2$ contingency test, we can
combine the medium and large categories, as shown in Table 9.6b.

Now there will be no problem with expected frequencies being too small.
They are all

$$\frac{column\ total}{grand\ total} \times row\ total = \frac{30}{60} \times 30 = 15$$

In passing we might note that some time is saved in using the "edge table"
representation here. If the initial data set looked like this

**Data Display**

```
Row         Treatment    Size   Plants
 1     No Fertiliser    Small      20
 2     No Fertiliser    Medium      7
 3     No Fertiliser    Large       3
 4   With Fertiliser    Small      10
 5   With Fertiliser    Medium     15
 6   With Fertiliser    Large       5
```

then we could easily aggregate medium and large by recoding the size column using **Data>Code>Text to Text**. If we call the new column "size 2," we would see this:

**Data Display**

| Row | Treatment | Size | Plants | Size 2 |
|-----|-----------|------|--------|--------|
| 1 | No Fertiliser | Small | 20 | Small |
| 2 | No Fertiliser | Medium | 7 | Large |
| 3 | No Fertiliser | Large | 3 | Large |
| 4 | With Fertiliser | Small | 10 | Small |
| 5 | With Fertiliser | Medium | 15 | Large |
| 6 | With Fertiliser | Large | 5 | Large |

The reason for ensuring that expected values are greater than 5 is that the test is oversensitive to small differences between observed and expected when the expected value is small. This is because dividing by a very small expected value (imagine dividing by 0.1) will give rise to a very high component of chi-squared for that cell, not justified by the weight of evidence.

## 9.4 A Further Example of a Chi-Squared Contingency Test

We have carried out an experiment to compare the effectiveness of three termite repellents in preserving fencing stakes. Each chemical was applied to a sample of 300 fencing stakes, giving 900 stakes in total. The usual precautions were taken to ensure that each stake represented an independent trial of the effectiveness of the repellents (see Chapter 4). The number of stakes that were attacked was recorded after a year in the ground.

```
MTB > print c10—c12
```

**Data Display**

| Row | Behaviour | Preservative | Stakes |
|-----|-----------|--------------|--------|
| 1 | Attack | Oil | 112 |
| 2 | Attack | Creosote | 82 |
| 3 | Attack | Cu arsenate | 123 |
| 4 | Avoid | Oil | 188 |
| 5 | Avoid | Creosote | 218 |
| 6 | Avoid | Cu arsenate | 177 |

```
Tabulated statistics: preservative, behaviour

Using frequencies in stakes
Rows: preservative Columns: behaviour

                Attack   Avoid    All
Creosote            82     218    300
                 105.7   194.3  300.0
                 5.301   2.882      *
Cu arsenate        123     177    300
                 105.7   194.3  300.0
                 2.843   1.546      *

Oil                112     188    300
                 105.7   194.3  300.0
                 0.380   0.206      *

All                317     583    900
                 317.0   583.0  900.0
                     *       *      *


Cell Contents: Count
               Expected count
               Contribution to chi-square

Pearson chi-square = 13.158, d.f. = 2, p-value = 0.001
Likelihood ratio chi-square = 13.394, d.f. = 2, p-value = 0.001
```

The table value (Table C.6) of $\chi^2$ for 2 d.f. at $p = 0.05$ is 5.99, and for $p = 0.01$ it is 9.21. With a calculated $\chi^2$ of 13.158, we have strong evidence to reject the Null hypothesis that all three chemicals are equally effective. MINITAB provides the value $p = 0.001$.

If we examine the components of $\chi^2$, printed below the expecteds in the table by selecting the "Each cells contribution … " under the **Chi-Square** Button of **Stat>Table>Cross tabulation** … we see that the two coming from creosote are especially high (5.301 and 2.882). This tells us that it is creosote that has a different effect from the others. Comparing the observed and expected values, we see that creosote was more effective than the other two treatments with only 82/ 300 = 27% of stakes being attacked, compared with 37% and 41% for the others.

## 9.5  Beyond Two-Dimensional Tables: The Likelihood Ratio Chi-Square

An alternative approach to the analysis of frequency data is exemplified by the **G-test**, which MINITAB prints at the bottom of a contingency analysis under the title of "Likelihood Ratio Chi-Squared test." Although in the simple cases we have dealt with in this chapter the two tests give very similar (though not always identical) results, the G-test is the tip of a very interesting iceberg of methods for dealing with more complex frequency tables.

The calculation of the G-test statistic, which you can check for yourself quite easily, is

$$2 \times \sum O \ln\left(\frac{O}{E}\right).$$

The mathematical function **ln()** is the **natural logarithm**, and the quantity

$$\sum O \ln\left(\frac{O}{E}\right)$$

is called the **log likelihood** for the observed frequencies O, given that the expected frequencies are E.

The reason why this is of interest is that it points the way towards more complex analyses involving more than two predictors. The key to understanding how this works is to see the analogy between a contingency test such as the fencing stake analysis in Section 9.4, and ANOVA (Chapter 6). The predictor variables are insect behaviour (attack/avoid) and treatment. The response variable, less obviously, is the number of stakes (i.e., their frequency) in each category. The "regular" chi-squared test, which we must now learn to call the Pearson chi-squared test, is a significance test of the interaction term behaviour*treatment. The "main effects" are the marginal frequencies, which tell us the overall proportion attacked and the overall proportion in each treatment, neither of much interest in themselves. Imagine, however, that we also wanted to include stakes made of different kinds of wood in our analysis. How could we find out whether the effectiveness of the different treatments depended on what the stake was made of? Although, in principle, we could extend the Pearson chi-square from two to three dimensions, this would not be very useful as the only hypothesis we could test in a single analysis would be the three-way interaction behaviour*treatment*wood. To find the behaviour*treatment interaction without regard to wood, we would have to collapse the table back into two dimensions, and as the number of factors increases, the number of such tests becomes unacceptably large.

The modern approach to this is to use the **log linear model**, which is an extension of the G-test (likelihood ratio chi-squared test) into more dimensions. If you need to use these methods, we suggest you consult a more advanced textbook such as Howell (2001) or Crawley (2005). Although it provides the G-test alongside the Pearson test, more complex Log Linear Models are not implemented in MINITAB. They are to be found in more advanced packages such as R, SPSS, SAS, and Genstat.

# 10

## Nonparametric Tests

It is vain to do with more what can be done with less.

—**William of Occam**

### 10.1 Introduction

Most of the statistical techniques we have considered so far are **parametric** tests (t-tests, ANOVA, and linear regression). This means that, in addition to the assumption that each data point is an independent test of the hypothesis in hand (one shared by all statistical tests), they depend upon the assumptions listed at the end of Chapter 6 and Chapter 8, namely,

- Normality of error. The residuals must follow the Normal distribution.
- Homogeneity of variance. We have seen three instances of this requirement, Equality of variances in a two sample t-test (Chapter 3, Section 3.8), homogeneity in ANOVA (Chapter 6, Section 6.10.3), and homogeneity in regression (Chapter 8, Section 8.8.3).
- Linearity (where appropriate). This has only been applied in linear regression so far (Chapter 8, Section 8.8.4).

There is also an additional assumption that we have not previously met, which is that where data are quantities, they must be measured at least on an **interval scale** for the relationship between test statistic and probability to work as advertised. An interval scale is one in which the intervals are meaningful. For example, in the Celsius scale the difference (interval) between 10° and 20° is the same as that between 20° and 30°. By contrast, the Beaufort wind scale is not an interval scale. Force 3 (gentle breeze) is more than Force 2 (light breeze), but the difference of one unit between Force 2 and Force 3 cannot be regarded as equivalent to the difference between a Force 9 (severe gale) and a Force 10 (storm). Thus, the Beaufort scale is a **rank scale** but not an interval scale.

If any of these assumptions fails, then we may have to use **a nonparametric test**, one which relies on less stringent assumptions. If we apply a parametric test when its assumptions are not valid, we obtain an incorrect p-value. This means that the probability of a Type I error may be larger than is apparent from the calculations.

### 10.1.1   When Are Assumptions Likely to Fail?

The Normality of error assumption can fail if, for example, residuals from the test show skew to the right or to the left (Figure 10.1a). We can also run

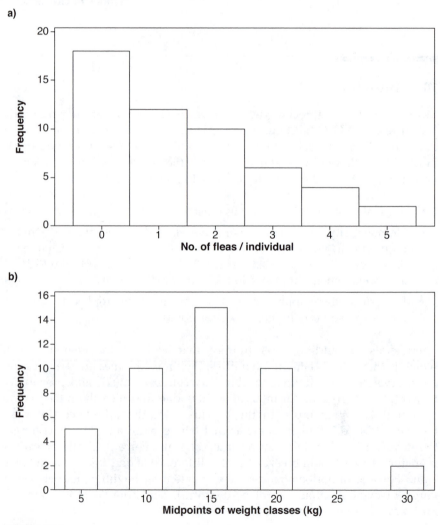

**FIGURE 10.1**
(a) Positive skewness; (b) outliers.

into problems if the sample sizes are small because the central limit theorem, which guarantees that the sampling distribution of parameters from any shaped population tend to be normal (Section 1.5.2), only does so if sample sizes are "large."

It is a common mistake to think that the *data* must be Normally distributed, but this is not necessarily the case. It is true that, with small samples, skewed data distributions may result in the error being skewed, but there are many circumstances in which raw data may be anything but Normal but the error distribution is perfectly acceptable. As an example, consider Figure 10.2a and b, which shows the distribution of heights of a group of students of both genders. If the model is

$$height = gender$$

then it is quite likely that the error distribution will be reasonably Normal (Figure 10.2c) even though the data are not.

Extreme observations (outliers, Figure 10.1b) may cause problems with Normality of error, or with homogeneity of variance, as we have seen in the Anscombe data set examples in Chapter 8, Section 8.9.

Nonlinearity may arise from use of an inappropriate scale of measurement. For example, if a fertiliser dose affects the linear dimensions of a plant but we actually measure the leaf area (which is proportional to the linear dimensions squared), we will probably find that the linearity assumptions of a parametric model are violated. It may be possible to modify the data so that

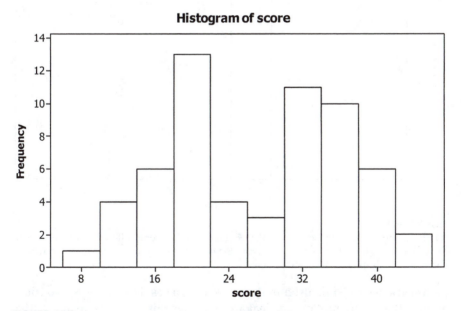

**FIGURE 10.2a**

(a) Histogram of scores without regard to any other variable. The distribution of the raw data is not normal.

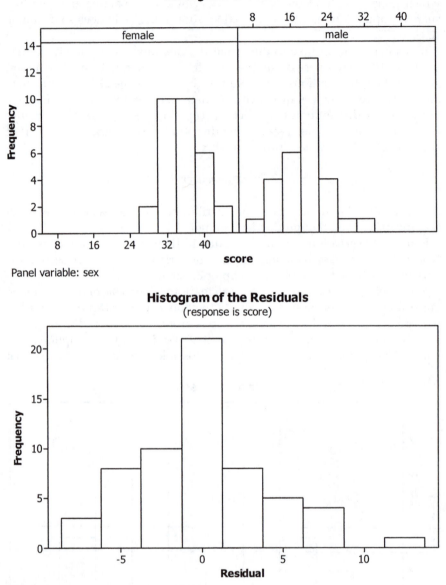

**FIGURE 10.2**
(b) Separating the data by sex shows that Figure 10.2a is a composite of two more or less Normal-looking distributions. (c) Residuals of the model *score = sex* are reasonably Normal, so parametric analysis is to be preferred here.

parametric tests can be used in these circumstances. For example, we could analyse the square root of each leaf area value as this provides an alternative measure of leaf size, which would probably be linear. This is called **trans-formation** (see Grafen and Hails, 2002). However, if we cannot find a suitable transformation or wish to analyse the raw data, we need to consider non-parametric tests. If the data are not measured on an interval scale, then there is no choice; we must use nonparametric methods.

## 10.2  Basic Ideas

A number of basic techniques are shared by all the tests in this chapter, so we group them here.

### 10.2.1  Ranking

Most nonparametric tests involve ranking the data. It does not really matter whether you rank from smallest to largest or largest to smallest, but it is usual to give the smallest value rank 1, so that the rank scale goes in the same direction as the original data. For $n$ observations, the

$$\text{sum of ranks} = \frac{n(n+1)}{2},$$

and the mean rank (which is the rank of the median observation) is

$$\frac{(n+1)}{2}.$$

This means that the *order* of the data is preserved but the actual relative size is discarded, which often removes distribution problems. Here are two examples:

**Data Display**

| Row | Scores1 | Rank1 |
|-----|---------|-------|
| 1 | 1.0 | 1 |
| 2 | 3.0 | 2 |
| 3 | 4.0 | 3 |
| 4 | 6.0 | 4 |
| 5 | 8.5 | 5 |

**Data Display**

| Row | Scores2 | Rank2 |
|-----|---------|-------|
| 1 | 4.2 | 1.0 |
| 2 | 5.0 | 2.5 |
| 3 | 5.0 | 2.5 |
| 4 | 6.0 | 5.0 |
| 5 | 6.0 | 5.0 |
| 6 | 6.0 | 5.0 |
| 7 | 7.0 | 7.0 |

The first example is unproblematic; the data were ordered so you can see what is going on, but this is not necessary for the RANK procedure to work. In the second example, both the scores with value 5 tie with a rank of 2.5, and the 3 scores with value 6 have the average rank of 5. The commonest error when doing this by hand is to forget that, in this case, there will be no rank 4 or rank 6. It is a good idea to add up the ranks and check that

$$ranksum = \frac{n(n+1)}{2}$$

For $n = 7$, the answer should be 28.

### 10.2.2  Measures of Centre and Scatter

The whole point about nonparametric tests is that we do not assume that the data are on an interval scale. This means that the arithmetic operations of addition, subtraction, multiplication, and division do not give sensible answers, so we cannot make use of the mean and the standard deviation because those operations are used in calculating them. Instead, we use statistics that are based on the ranks of the data rather than the original measurements.

### 10.2.3  Median

The preferred measure of centre is the median (or the "middle number"), which is the measurement that has rank

$$\frac{(n+1)}{2}.$$

In the case of "scores1" in the example above, this is easy: the rank of the median is

$$\frac{5+1}{2} = 3,$$

and the third-ranking score is 4.0. Where there are ties, things can get more complicated. MINITAB says that the median for "scores2" is 6, which is the number in the

$$\frac{7+1}{2} = 4th \text{ row from the top.}$$

If we had an even number of scores, the median would be the one ranking

$$\frac{(6+1)}{2} = 3.5,$$

which we would take to be halfway between the third- and fourth-ranking observation.

Getting the median for grouped data can be very tedious. Rees (2001) gives a graphical method.

## 10.2.4. Quartiles and Other Quantiles

The measure of scatter that we usually use in nonparametric statistics is the inter quartile range. The lower quartile is the value such that 25% of the data fall below it. Its rank would be

$$\frac{(n+1)}{4}.$$

In "scores2," this is the

$$\frac{(7+1)}{4} = 2\text{nd ranking observation (5.0)}.$$

For "scores1", we have a problem in that the lower quartile has a fractional rank

$$\frac{(5+1)}{4} = 1.5,$$

so we take the value to be 0.5 of the interval between the observation having rank 1 (1.0) and that having rank 2 (3.0), to get a value of 2.0.

Similarly, the upper quartile is the value such that 25% of the data fall above it, and its rank is

$$(n+1) \times \frac{3}{4} = (n+1) \times 0.75.$$

Applying this formula for "scores1" gives us a rank of 4.5 for the upper quartile, which is defined as lying at 0.5 of the interval between the 4th rank score (6.0) and the 5th ranking score (8.5) to give 7.25.

The interquartile range, therefore, is the range within which half the data lie. These quantities are all shown below:

```
Descriptive Statistics: Scores1, Scores2

Variable   N   N*   Minimum      Q1   Median      Q3   Maximum      IQR
Scores1    5    0      1.00    2.00     4.00    7.25      8.50     5.25
Scores2    7    0     4.200   5.000    6.000   6.000     7.000    1.000
```

We might choose any percentile value to be of interest under some circumstances. For example, the 5th percentile value (the value such that 5% of the observations fall below it) would be the observation with rank $(n + 1) \times 0.05$. We can sometimes deal with outliers in otherwise reasonable data by removing the upper and lower 5th or 10th percentiles, working with **truncated** data.

## 10.3   A Taxonomy of Tests

In this chapter we will discover that each of the parametric techniques we have learned has a nonparametric cousin that can be used when the assumptions needed for the parametric version are not justifiable. Before we start, let us recall the basic methods of testing procedure, which apply just as much here as in the rest of the book:

### 10.3.1   Hypothesis Testing

- Develop *research hypothesis* $H_A$.
- Deduce *Null hypothesis* $H_0$, which we will try to disprove.
- Collect data: from it calculate a *test statistic*.
- Use distribution of test statistic to find the probability of calculated value *if $H_0$ is *true*.
- Decide on criterion probability $\alpha$ for rejection of $H_0$.

**Decision**: Reject $H_0$ if the probability of calculated value less than $\alpha$.

### 10.3.2   Test Statistics

We have become familiar with a number of **test statistics** in previous chapters: z, t, F, r, $\chi^2$, to name but five. These are quantities *calculated from the data* whose sampling distribution is known *if the Null hypothesis is true*. One confusing thing about nonparametric tests is that almost every test has its own special test statistic. Most of them depend on the properties of the *Uniform distribution*; some on the *Binomial distribution*. When doing these tests by hand, note:

- For small samples (generally), a special set of tables or a computer may be needed for the relevant test statistic, but
- For large samples, there is generally an approximation to a parametric test statistic we already know about.

Some of these approximations are quite complicated to calculate and will not be covered here. Let the machine take the strain!

We recommend Hollander and Wolfe (1999) for a complete and authoritative treatment of nonparametric tests.

## 10.4  Single-Sample Tests

In Chapter 2 we considered the problem of deciding whether a sample of 30 catches of Patagonian toothfish provided evidence that the mean tonnage of fish landed in a week would be more than 40 t; the reason we want to know this is that we believe that the fishery would not be economically viable if the mean catch was too small. A formal test for Normality in the toothfish data set reveals that they do depart slightly from the Normal distribution, so there might be reservations about using the $z$ or $t$ test as we did in Chapter 2. We could also suppose, just for illustration, that the same numbers actually represented a "subjective" score assigned to 30 experimental subjects in a psychometric test. In this alternative example, we believe that the ordering of the scores is correct, but we are not confident that the scale is truly an interval scale. There are two ways to address this.

### 10.4.1  Sign Test

The sign test is very conservative, by which we mean that it is unlikely to lead us astray through a Type I error, but it may lack statistical power because a Type II error is perhaps more likely that we would choose. The logic of the test is

- Research hypothesis: the median score is more than 40, so more scores will be above 40 than below.
- Null hypothesis: the median score is 40 or less, so scores will lie equally above and below 40, or more will be below.
- Test statistic: the number of scores below the median. You can see from Figure 10.3 that 8 out of the 30 scores lie below 40, whereas 3 are exactly equal to 40.

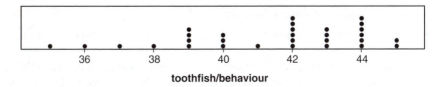

**FIGURE 10.3**
Dot plot of scores assigned to 30 subjects in a psychometric test.

This test statistic follows the binomial distribution. We are asking the equivalent to "What is the probability of tossing a fair coin 27 times and getting 8 heads?" It is 27 and not 30 because scores equal to the test median are considered to give us no information and are therefore discarded. MINITAB implements the sign test as **Stat>NonParametrics>1-Sample Sign test**.

```
Sign Test for Median: toothfish/behaviour

Sign test of median = 40.00 vs. > 40.00

                        N  Below  Equal  Above       p  Median
Toothfish/behaviour    30      8      3     19  0.0261   42.00
```

The probability of getting as few as 8 scores below the median if the Null hypothesis (median is 40 or less) is true is 0.0261, so we can reject the Null hypothesis and accept the alternative that the median is > 40.

Note that this is a *one-tailed test* because the alternative hypothesis is that the score is > 40; a two-tailed test would be the case where the alternative hypothesis is that the score is not equal to 40, as follows:

```
Sign Test for Median: toothfish/behaviour

Sign test of median = 40.00 vs. not = 40.00

                        N  Below  Equal  Above       p  Median
Toothfish/behaviour    30      8      3     19  0.0522   42.00
```

The reasoning is the same as we outlined in Section 2.5; the result of the two-tailed test looks almost the same as the one-tailed example, but the p-value is exactly twice as big. Here, if the test was two tailed we would not reject $H_0$.

### 10.4.2  Wilcoxon Single-Sample Test

The sign test is extremely conservative because it uses only the direction of each score relative to the Null hypothesis median. A more powerful version is the Wilcoxon test. In this test we subtract the Null hypothesis median from each score. We then rank the differences without regard to their sign, and finally we add together the ranks for the positive differences and those for the negative differences separately. For simplicity we will do this test two-tailed:

- Research hypothesis: the median score is not equal to 40.
- Null hypothesis: the median score is 40.
- Test statistic: the sum of the ranks of scores that are above 40.

The test statistic (Wilcoxon statistic, W) requires special tables, for which a specialist text on nonparametric methods should be consulted. You can see intuitively that if the Null hypothesis is true, the sum of the ranks of scores above the Null hypothesis median should be about equal to the sum of ranks of scores below it, with half of the scores lying either side. Thus, the expected value of W would be

$$\frac{n(n+1)}{4}.$$

However, the theory predicting the sampling distribution for W is complicated, and in practice, we rely on tables. MINITAB provides a p-value in the usual way.

```
Wilcoxon Signed Rank Test: toothfish/behaviour

Test of median = 40.00 vs. median not = 40.00

                            N for    Wilcoxon              Estimated
                       N    Test    Statistic      p        Median
Toothfish/behaviour   30     27       295.0     0.011       41.50
```

This example illustrates the fact that the Wilcoxon test is more powerful than the sign test when applied to the same data. Here, we would reject $H_0$ ($p < 0.05$) and conclude that the median weekly catch was not 40 t (in fact, it is larger: median 41.5 t). Using the less powerful two-tailed sign test in the previous section did not allow us to reject $H_0$ ($p = 0.0522 > 0.05$).

## 10.5 Matched-Pairs Tests

In Chapter 3 we considered seven pairs of twin sheep where one of each pair was raised on a new diet. We want to know whether their growth is different from the other member of the pair, raised on the standard diet. Here are the data; we have also calculated the difference *new–old* for each pair. Let us now suppose that instead of weight, the numbers represent a subjective score of the wool condition after 10 weeks, and that this can be regarded only as a rank scale.

```
Data Display

Row   Old   New   Diff
  1   3.4   4.5   1.1
  2   3.9   4.8   0.9
  3   4.2   5.7   1.5
  4   4.5   5.9   1.4
  5   3.6   4.3   0.7
  6   2.9   3.6   0.7
  7   3.2   4.2   1.0
```

### 10.5.1    Sign Test and Wilcoxon on Differences

If both diets had the same effect on the sheep, then we would expect the median difference between them to be 0, so the logic of the test is

- Research hypothesis: the median difference is not zero.
- Null hypothesis: the median difference is zero.
- Test statistic: number of differences above and below the expected median of 0.

So, the test is just a single sample sign test on the differences between members of matched pairs. As we do not know in which direction to expect an effect, we conduct this test two-tailed.

```
Sign Test for Median: diff

Sign test of median = 0.00000 vs. not = 0.00000

       N  Below  Equal  Above      p  Median
Diff   7    0      0      7   0.0156   1.000
```

We reject the Null hypothesis (p < 0.05) and conclude that the diets do differ in their effect. The estimated difference in the medians is +1, so, as the difference was calculated as *new–old*, we conclude that the sheep on the new diet have better wool.

The same argument applies for the more powerful Wilcoxon test. We just apply the test to the differences, with Null hypothesis that the median of the differences is 0.

```
Wilcoxon Signed Rank Test: Diff

Test of median = 0.000000 vs. median not = 0.000000

                N for
            Wilcoxon  Estimated
       N      Test    Statistic       p  Median
Diff   7       7        28.0      0.022   1.050
```

Here we reject $H_0$ (W = 28, p = 0.022) and again conclude that the diets have an effect on wool quality, with the new diet group having higher scores (estimated median difference = 1.05).

## 10.6   Independent Samples

### 10.6.1    Two Groups: Mann–Whitney Test

If we have an experiment with two treatments in which our data do not meet the requirements for a parametric test, we should use a Mann–Whitney test (sometimes known as the **Wilcoxon Rank Sum** test). This test shows whether there is a difference between two population medians. The assumptions we must be able to make are

- The observations must be random and independent observations from the populations of interest (as with all statistical tests).
- The samples are assumed to come from populations that have a similar shaped distribution (for example, both are positively skewed).

The growth form of young trees grown in pots can be scored on a scale from 1 (poor) to 10 (perfect). This score is a complex matter as it summarises the height, girth, and straightness of the tree. We have 15 trees. Of these, 8 trees were chosen at random to be grown in a new compost, whereas the remaining 7 were grown in a traditional compost. After 6 months' growth the scores were recorded:

**Data Display**

| Row | Newcomp | Oldcomp |
|-----|---------|---------|
| 1 | 9 | 5 |
| 2 | 8 | 6 |
| 3 | 6 | 4 |
| 4 | 8 | 6 |
| 5 | 7 | 5 |
| 6 | 6 | 7 |
| 7 | 8 | 6 |

So, we have

- Research hypothesis: the composts differ in their effect.
- Null hypothesis: the two composts produce trees with the same median growth form.

MINITAB carries out the Mann–Whitney test on two columns as **Stat>NoParametrics>Mann–Whitney First sample** newcomp, **Second sample** oldcomp.

**Mann–Whitney Test and CI: newcomp, oldcomp**

```
             N   Median
newcomp      8    7.500
oldcomp      7    6.000

Point estimate for ETA1-ETA2 is 2.000.
95.7% CI for ETA1-ETA2 is (1.000, 3.000).
W = 86.0.
Test of ETA1 = ETA2 vs. ETA1 not = ETA2 is significant at 0.0128.
The test is significant at 0.0106 (adjusted for ties).
```

First, the number of observations (N) and the median for each treatment is given, followed by an estimate of the difference between the two medians (ETA1–ETA2), together with a confidence interval for this difference. ETA stands for the Greek letter $\eta$, which is the conventional symbol for the population median. This seems rather odd at first sight, as the difference between the estimated medians (7.5–6.0.) appears to be 1.5. However, the Mann–Whitney "point estimate" for the difference is made using a more sophisticated calculation. It calculates the median of all the pairwise differences between the observations in the two treatments. This gives a value of 2.0.

The confidence interval (1.0–3.0) is at 95.7%. This is the range of values for which the Null hypothesis is not rejected, and is calculated for a confidence level as close as possible to 95% (again, the method involves repetitive calculations that can only sensibly be carried out by a computer program).

Finally, the test statistic W is calculated. All the scores are ranked regardless of group. Then the sum of the ranks of the groups are calculated separately:

| Score | 4 | 5 | 5 | 6 | 6 | 6 | 6 | 6 | 6 | 7 | 7 | 7 | 8 | 8 | 9 |
|---|---|---|---|---|---|---|---|---|---|---|---|---|---|---|---|
| Compost | O | O | O | O | O | O | N | N | O | N | N | N | N | N | N |
| Rank | 1 | 2.5 | 2.5 | 6 | 6 | 6 | 6 | 6 | 10 | 10 | 10 | 13 | 13 | 13 | 15 |

The sum of the ranks for the new compost trees (the first group) is

$$W = 6 + 6 + 10 + 10 + 13 + 13 + 13 + 15 = 86$$

Then the p-value for the test is given (p = 0.0128): This calculation assumes that there are no tied values in the data. As there are some ties in this case, an adjustment has to be made to the calculation of the probability level to give p = 0.0106.

We can reject $H_0$ and conclude that the two composts differ in their effect on tree growth form. The new compost is preferable as it produces trees with a higher median score.

### 10.6.2   More than Two Groups: Kruskal–Wallis Test

Suppose that we have data from an experiment with more than two treatments. We want to ask questions of it as we would in ANOVA, but we find that the data do not satisfy the requirements for parametric tests. For example, we find that the observations cannot be regarded as measurements on an interval scale. The nonparametric equivalent of the fully randomised or one-way ANOVA is the **Kruskal–Wallis test**. It can be considered a multiple group generalisation of the Mann–Whitney test, and as such, it makes the same assumptions: first, the observations must be random and independent observations from the populations of interest. Second, the samples we compare are assumed to come from populations that have a similar shaped distribution. This does not have to be Normal. It could be that both or all tend to have a few large values and so have "positive skew" (Figure 10.1a), or they could both or all be negatively skewed. Whatever the shape, they must both or all share it.

We will illustrate the use of this test on two data sets. First, we will reexamine the data we analysed in Chapter 6, using one-way ANOVA. To remind you, the data are from four treatments (types of vegetation management by sowing and cutting), each replicated four times on plots whose positions were randomly selected around a field margin. The response variable is the number of spiders per plot.

**Data Display**

| Row | Treatment | Spider |
|-----|-----------|--------|
| 1 | F1 | 21 |
| 2 | F1 | 20 |
| 3 | F1 | 19 |
| 4 | F1 | 18 |
| 5 | F2 | 16 |
| 6 | F2 | 16 |
| 7 | F2 | 14 |
| 8 | F2 | 14 |
| 9 | NF1 | 18 |
| 10 | NF1 | 17 |
| 11 | NF1 | 15 |
| 12 | NF1 | 16 |
| 13 | NF2 | 14 |
| 14 | NF2 | 13 |
| 15 | NF2 | 13 |
| 16 | NF2 | 12 |

MINITAB carries out the Kruskal–Wallis test using **Stat>Nonparametrics> Kruskal–Wallis Response**: spider **Factor:** treatment. Note that here the data must be in an "edge table," unlike the separate column format used for the Mann–Whitney test. (See Appendix A, Section A.7.)

**Kruskal–Wallis Test: Spider vs. Treatment**

Kruskal–Wallis Test on Spider:

| Treatment | N | Median | Ave Rank | z |
|-----------|-----|--------|----------|-------|
| F1 | 4 | 19.50 | 14.4 | 2.85 |
| F2 | 4 | 15.00 | 7.0 | −0.73 |
| NF1 | 4 | 16.50 | 9.9 | 0.67 |
| NF2 | 4 | 13.00 | 2.8 | −2.79 |
| Overall | 16 | | 8.5 | |

H = 12.66  d.f. = 3  p = 0.005
H = 12.85  d.f. = 3  p = 0.005 (adjusted for ties)

NOTE * One or more small samples

First, the number of observations in each treatment is presented (N) together with their median values. Then, all 16 observations are ranked, and the mean rank for each treatment is given. Finally, a "z-value" is calculated for each treatment. This shows how the mean rank for each treatment differs from the mean rank for all 16 observations, and details of its calculation are given in Box 10.1.

---

## Box 10.1    Calculation of z-Scores in the Kruskal–Wallis Test

For convenience, the mean rank for all observations has been converted to zero. This overall mean has been subtracted from each treatment's mean rank, and the result has been divided by the standard deviation of that treatment's ranks to give a z-value for each treatment. These show the location of each treatment's mean rank around zero (Figure 10.4) — the overall mean of this standard normal distribution. We remember from Chapter 2 that 95% of the values would be expected to lie between +1.96 and –1.96 units from the mean. Here we see that two treatments have z-values outside this range, suggesting that not all treatments are likely to come from one population.

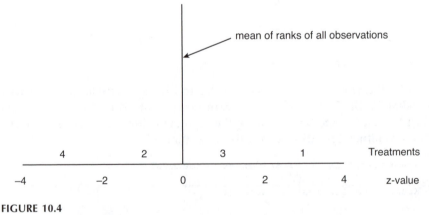

**FIGURE 10.4**
Calculation of z-values for the Kruskal–Wallis test.

---

The test statistic H is printed with its degrees of freedom (*number of groups* – 1, just as in ANOVA) and a p-value. H, in fact, follows the $\chi^2$ distribution should you ever have to conduct this test by hand. We can reject the Null hypothesis that the observations all come from the same population (because the value of p is very much less than 0.05). We have strong evidence that at least one of the four treatments differs from at least one of the other four in terms of its spider population.

A second version of the test statistic, adjusted for ties, H, is also given. This differs slightly from the first in that an adjustment has been made to account for the presence of tied observations (for example, there are several plots with 16 spiders). This has to be done because the test assumes that the observations come from a continuous distribution (which could contain values like 2.13459), whereas we have used it to analyse counts where ties are more likely to occur. We should use this second, corrected version of the statistic.

*Small sample warning:* In the computer printout we see a warning about small samples. The problem here is that MINITAB uses the assumption that the Kruskal–Wallis test statistic H follows the $\chi^2$ distribution, but this approximation is only reliable when the groups all have five or more observations. Some caution is needed in interpreting an analysis with fewer than this. There are two ways to deal with this problem:

1. Seek out tables of H for small sample cases. Hollander and Wolfe (1999) and Gibbons (1993a) have tables for up to three groups.
2. Find a way to calculate the probability of the observed H.

One possibility at the time of writing is to use the freeware program Exact Test (see http://www.exact-test.com/exactstat.html) Using this program with the spider data set gives an exact probability very close to that calculated by MINITAB from the $\chi^2$ distribution, so, in this case, we can accept the analysis at its face value in spite of the warning. Perhaps the best advice would be to plan to collect sufficient data in the first place (we included this analysis to show the similarities with the parametric ANOVA of the same data in Chapter 6).

### 10.6.2.1 Another Example of the Kruskal–Wallis Test

A student has scored the amount of bacterial growth on each of 20 petri dishes. A score of 0 indicates no growth, whereas one of 5 shows that the bacterium covers the entire dish. Five days previously, 10 randomly selected petri dishes had received a standard amount of an established antibiotic used to inhibit bacterial growth (treatment 1), whereas the remaining 10 dishes had received the same amount of a newly discovered inhibitory compound (treatment 2).

```
Data Display

Row    Antibiotic    Score
  1    established      1
  2    established      2
  3    established      3
  4    established      2
  5    established      3
  6    established      4
```

```
 7  established     2
 8  established     1
 9  established     2
10  established     3
11        new       3
12        new       3
13        new       5
14        new       4
15        new       2
16        new       5
17        new       4
18        new       3
19        new       4
20        new       3
```

A preliminary plot of the data suggests that the new compound may not be much of an improvement (Figure 10.5); if anything, it is not as good as the established one.

The Kruskal–Wallis test tests the Null hypothesis of no difference between the two populations.

```
MTB > Kruskal–Wallis "score" "antibiotic."
```

**Kruskal–Wallis Test: Score vs. Antibiotic**

```
Kruskal–Wallis test on score

Antibiotic     N  Median  Ave Rank       Z
Established    10   2.000       7.3   -2.46
       New     10   3.500      13.8    2.46
   Overall     20               10.5

H = 6.04   d.f. = 1   p = 0.014
H = 6.46   d.f. = 1   p = 0.011 (adjusted for ties)
```

We have evidence to reject the Null hypothesis ($p < 0.05$). We conclude that the new chemical does differ from the established product; it is, in fact, a worse inhibitor of bacterial growth.

### 10.6.3   Several Groups with Blocking: Friedman's Test

In ANOVA (Chapter 6) we saw that blocking was a useful way of taking account of known sources of variation that might interfere with our results, thus making it easier to detect the effect of treatments (if there are any). Friedman's test allows us to do the same thing with data that do not satisfy the requirements of parametric tests. Like both the Mann–Whitney and

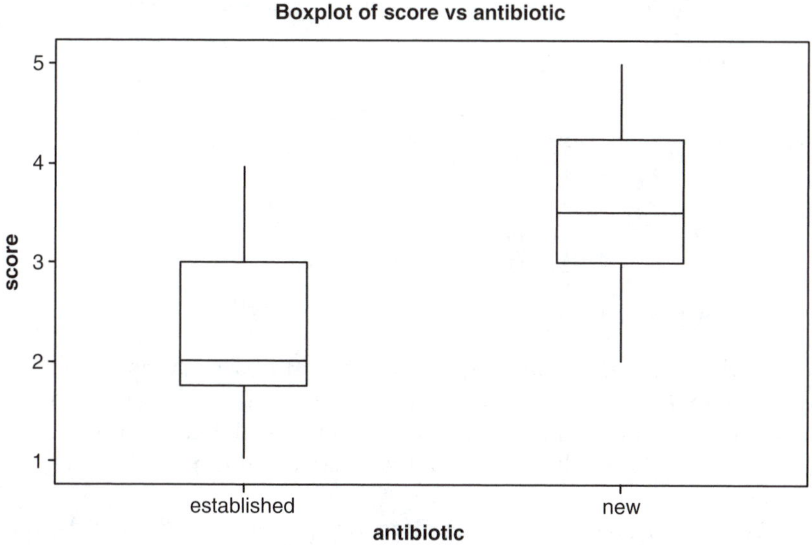

**Boxplot of score vs antibiotic**

**FIGURE 10.5**

Box plot of scores of bacterial growth on petri dishes receiving an established and a new antibiotic.

Kruskal–Wallis tests, Friedman's test does assume that the samples come from populations with similarly shaped distributions.

In the field experiment on the effects of sowing and cutting management on spider numbers, we allocated one of each treatment to each of four blocks, with a block being one side of the field.

**Data Display**

| Row | Treatment | Spider | Block |
|-----|-----------|--------|-------|
| 1 | F1 | 21 | 1 |
| 2 | F1 | 20 | 2 |
| 3 | F1 | 19 | 3 |
| 4 | F1 | 18 | 4 |
| 5 | F2 | 16 | 1 |
| 6 | F2 | 16 | 2 |
| 7 | F2 | 14 | 3 |
| 8 | F2 | 14 | 4 |
| 9 | NF1 | 18 | 1 |
| 10 | NF1 | 17 | 2 |
| 11 | NF1 | 15 | 3 |
| 12 | NF1 | 16 | 4 |
| 13 | NF2 | 14 | 1 |
| 14 | NF2 | 13 | 2 |
| 15 | NF2 | 13 | 3 |
| 16 | NF2 | 12 | 4 |

```
Friedman's Test: Spider vs. Treatment Blocked by Block

S = 12.00 DF = 3 P = 0.007

                                    Sum of
    Treatment   N  Est Median     Ranks
           F1   4       19.344      16.0
           F2   4       14.844       8.0
          NF1   4       16.469      12.0
          NF2   4       12.719       4.0

Grand median = 15.844
```

As with the Kruskal–Wallis test, the number of observations in each treatment is given together with their median values. Here the sum of the ranks of each observation within each block is then calculated. So, the sum of 16 for treatment 1 shows that it received the maximum rank of 4 in each of the 4 blocks.

Above this summary lies the test statistic, S, together with the degrees of freedom (4 treatments − 1 = 3) and the p-value (0.008). This is slightly smaller than the p-value (0.014) obtained by ignoring blocking (Kruskal–Wallis, Section 10.6.2). In other words, as we might expect in a well-designed experiment, blocking has caused an increase (albeit small in this case) in the efficiency of detecting differences between treatments. Like Kruskal–Wallis H, Friedman's S follows the $\chi^2$ distribution if the Null hypothesis is true.

It is important to realise that we CANNOT use Friedman's test to break down the treatment effect into a separate CUT and FLOWERS effect, as we were able to with ANOVA. Neither can we assess the benefit of the blocking explicitly; we can only deduce that it must have been effective from the fact that the p-value is smaller with Friedman's test than it was for the Kruskal–Wallis test with the same data set.

*Small sample warning:* Strictly, the "small sample" problem described in Section 10.6.2 for the Kruskal–Wallis test also applies here. With groups of fewer than 5 observations, the use of the $\chi^2$ distribution as an approximation to Friedman's test statistic S must be treated with caution. As with the Kruskal–Wallis test, the solution is to seek out tables of the exact probability for S, or to use an exact calculation. The program Exact Test estimates the p-value here as 0.0001 compared with 0.007 for the $\chi^2$ approximation, so, in this instance, the approximation is actually slightly conservative. Clearly, some judgment is called for here. As a rule of thumb, we would recommend that if the p-value for the test is as low as 0.01, you would be fairly safe to reject the Null hypothesis on the basis of the $\chi^2$ approximation even when some samples are a little too small.

### 10.6.3.1 Antibiotic Experiment with Blocking

Returning to our second example of the effect of two chemicals on bacterial growth (Section 10.6.2.1), we can see that a Friedman's test could also be useful here. Suppose that it takes about 30 sec to score a petri dish for bacterial growth. Those that are last to be scored will have had longer to grow than those scored first. Such a difference could be important if, for example, all the treatment 1 dishes were scored first, then all the treatment 2 ones. It would be much better to arrange our dishes in 10 pairs, each containing one of each treatment. We then score each pair, randomly choosing whether to score treatment 1 or treatment 2 first within each pair, taking time of assessment into account in the analysis as a blocking effect. The two dishes scored first are given a block value of 1, whereas the last pair are each given a value of 10. This will remove any differences in growth due to time of assessment from the analysis.

**Data Display**

| Row | Antibiotic | Score | Time |
|-----|------------|-------|------|
| 1 | Established | 1 | 1 |
| 2 | Established | 2 | 2 |
| 3 | Established | 3 | 3 |
| 4 | Established | 2 | 4 |
| 5 | Established | 3 | 5 |
| 6 | Established | 4 | 6 |
| 7 | Established | 2 | 7 |
| 8 | Established | 1 | 8 |
| 9 | Established | 2 | 9 |
| 10 | Established | 3 | 10 |
| 11 | New | 3 | 1 |
| 12 | New | 3 | 2 |
| 13 | New | 5 | 3 |
| 14 | New | 4 | 4 |
| 15 | New | 2 | 5 |
| 16 | New | 5 | 6 |
| 17 | New | 4 | 7 |
| 18 | New | 3 | 8 |
| 19 | New | 4 | 9 |
| 20 | New | 3 | 10 |

**Friedman's Test: Score vs. Antibiotic Blocked by Time**

```
S = 4.90   d.f. = 1   p = 0.027
S = 5.44   d.f. = 1   p = 0.020 (adjusted for ties)
```

|  |  | Est | Sum of |
|---|---|---|---|
| Antibiotic | N | Median | Ranks |
| Established | 10 | 2.000 | 11.5 |
| New | 10 | 4.000 | 18.5 |

Grand median = 3.000

Here, the test statistic, S (as for the H statistic in the Kruskal–Wallis test) is still significant ($p < 0.05$), but the p-value has slightly increased compared to that from the Kruskal–Wallis test. It appears that blocking by time of assessment has not improved the precision of the experiment here. After all, with only 20 plates in total, the assessment should be complete in 10 min. However, if the bacteria grew very fast, and if there were 100 plates to assess, taking at least 100 min, then blocking for time could be very important.

## 10.7  Two Quantitative Variables: Spearman's Rank Correlation

If the assumptions for parametric tests are met by both variables, then we use the Pearson correlation coefficient (Chapter 8). If the data are not suitable for parametric analysis, then we use a nonparametric equivalent: **Spearman's rank correlation coefficient** $\rho$ (pronounced "rho"). In Figure 10.6a we see a scatter plot of the data. It is the same as used in Chapter 8 but with an additional outlying point which, as we saw in the discussion of the Anscombe data sets, may have a very large influence on the result. The parametric Pearson correlation for the data with the additional point is:

```
Correlations: A, B
Pearson's correlation of A and B = 0.781
p-Value = 0.001
```

Spearman's $\rho$ is calculated using the ranks of the observations as follows. The lowest value for A is 3.0, so this gets a rank of 1, whereas the highest value is 46.1, so this gets a value of 15. The amounts of B are also ranked, with 3.1 receiving a rank of 1 and 15.2 a rank of 15. MINITAB can do this for us using **Data>Rank**, putting the results in columns "rA" and "rB":

```
Data Display
Row      A      B   rA   rB
  1   46.1   14.0   15   14
  2   23.6   15.2   13   15
  3   23.7   12.3   14   13
  4    7.0   10.9    7   12
  5   12.3   10.8   11   11
  6   14.2    9.9   12   10
  7    7.4    8.3    9    9
  8    3.0    7.2    1    8
  9    7.2    6.6    8    7
 10   10.6    5.8   10    6
 11    3.7    5.7    4    5
 12    3.4    5.6    2    4
 13    4.3    4.2    5    3
 14    3.6    3.9    3    2
 15    5.4    3.1    6    1
```

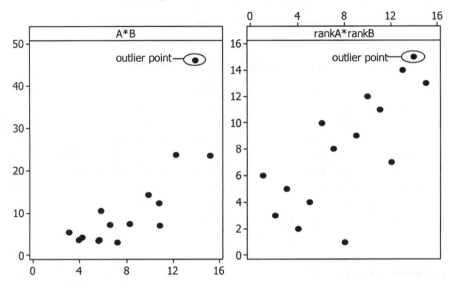

**FIGURE 10.6**
Scatterplot of (left) A versus B (right) rank of A versus rank of B. The data are the concentrations of two chemicals A and B in the same patient, as used in Figure 8.1 but with an additional outlier point added.

The correlation procedure is invoked in exactly the same way as for Pearson's correlation (**Stat>Basic Statistics>Correlation**); the only difference is that we specify the columns containing the ranks, not those with the original data.

```
Correlations: rA, rB
Pearson's correlation of rA and rB = 0.757
p-Value = 0.001
```

Note that MINITAB has no way of knowing that you are correlating ranks, so it misleadingly prints the label "Pearson's correlation." You should ignore this: the value is Spearman's correlation. The outlier clearly had some effect on the correlation coefficient because the "r" value for the ranked data is 0.757, compared with 0.781 for the original data. We can also see the difference by plotting the scatter plot of the ranked data (Figure 10.6b) to see whether there is a tendency for the two components of the matched pairs to increase and decrease together (or for one to increase as the other decreases). Ranking the data maintains the order of the observations on each axis. However, the size of the difference between one observation and the next biggest one is standardised. Now, the patient who was an outlier is still the point at the top right of the graph but, because the data are now ranked, he or she is not so far removed from the other data points as before. Spearman's $\rho$ is described as being more robust because it is less sensitive than Pearson's $r$ to occasional outliers or to bends in the relationship. However, the most compelling reason for using the nonparametric Spearman's $\rho$ rather than the parametric Pearson's $r$ would be the fact that one set or both sets of measurements could be considered only a rank scale.

### 10.7.1    Tied Observations in Spearman's Correlation

If there are a large number of observations having the same rank in either the x-variable or the y-variable then, in theory, the correlation coefficient obtained by just calculating Pearson's correlation of the ranks may produce a value that is too high. The MINITAB procedure outlined earlier ignores this problem on the grounds that effect is extremely small unless nearly all the observations are ties. You can probably ignore the problem, too, unless your result is close to the criterion for significance, *and* there is a high proportion of ties in the data, in which case consult a specialist text (e.g., Hollander and Wolfe, 1999) for a manual method of calculation.

## 10.8    Why Bother with Parametric Tests?

No, this is not a misprint! If a nonparametric test exists for every one of the tests we introduced earlier in the book, and if such tests make much less stringent assumptions about the data, why not just do everything with nonparametric methods? There are two main reasons for preferring parametric methods *when their assumptions are justified.*

### 10.8.1    Nonparametrics Have Lower Power Efficiency

To achieve the same **power** (probability of rejecting $H_0$ when it is, in fact, false), nonparametric tests need more data. This is not surprising, given that we are not using the information about the intervals between the measurements, only their rank order. The tests discussed here mostly have a power efficiency of about 90%, although provided their limited distributional assumptions are satisfied, the efficiency of the rank sum family of tests (Wilcoxon, Mann–Whitney, Kruskal–Wallis, and Friedman's tests) approaches 95%.

### 10.8.2    Available Nonparametric Analyses Are Less Complex

For example, the two-way ANOVA on spider numbers, where we decomposed the treatment into two main effects and an interaction term, has no nonparametric equivalent. Thus, with the exception of the limited blocked design available using Friedman's test, it is difficult or even impossible to use nonparametric tests to analyse complicated experimental designs and data where there are "nuisance" factors that need to be taken into account.

# 11

## Managing Your Project

The first 90% of a project takes 90% of the time; the last 10% takes the other 90%.

**—Anonymous**

Most undergraduate courses include a project and report writing component. Although experimental design, statistical analysis, and presentation of results have been covered already, this chapter brings together elements of project design and completion. Start thinking about your project early; it will take much more time than you think and probably carries a lot of marks. Beyond being satisfying, completing a good project develops skills that employers look for and contributes a large part to your final degree.

## 11.1 Choosing a Topic and a Supervisor

Whether or not you have an idea for a project, get help early. Your supervisor will have the biggest impact on your project, so choose someone who is appropriate (knows the subject area well) and is able to see you regularly. This may prevent a disaster. Ask previous students about who might be a helpful supervisor.

Your potential supervisor will probably be a very busy person. Make appointments, which should ensure his or her full attention. Have a well-prepared list of points ready to discuss. Open-ended questions are likely to generate open-ended answers. Ideally, advice should be sought at several stages throughout the project, so ask if you will be able to meet with the supervisor for regular discussion. If not, is there someone else who can help you should you need advice when your supervisor is necessarily absent? Pay attention to any deadline for agreeing on a project title and outline with your supervisor.

Your supervisor should have sufficient knowledge of the nature of your study to be able to recommend a satisfactory method of obtaining and analysing data. However, if there are any doubts, you should clarify your proposed methodology with a statistician.

When discussing your plans with your supervisor or a statistician, aim to produce a one- or two-paragraph summary of your research, including sample sizes and treatments, or a diagram of the layout. The important part is the experimental design and sampling protocol, as this summary will usually highlight statistical problems and show you where you have not thought matters through.

## 11.2   Common Mistakes

Most people only come to grips with statistics when they have their own data. This is why so many projects have major design flaws. However unattractive it may be, it is important to spend time working out how you will analyse your data before you collect it. In Chapter 4 we showed you how to create dummy data to simulate your experiment and analyse it before you carry it out. If you can, show your design to a statistician. If you do not do this, you will collect a large amount of data and then have to work out what to do with them. Many people do this and end up with one of the following:

1. A satisfactory analysis is possible that yields reasonable answers to the questions asked (**very rarely the case**).

2. All is well, but a more efficient design could have been used and could have saved a great deal of work (**very common**).

3. Analysis is possible, but the experiment was not sensitive enough to answer the question with sufficient precision. Perhaps, more replicates were required or extra measurements should have been taken. Another common problem here is pseudoreplication, where your replicates are not independent so that you have fewer replicates than you thought (**very common**).

4. The project is a disaster because the experimental design was badly flawed. Either no inferences can be made, or the results are not relevant to the required questions. In other words, your work was a waste of time and your project is a failure (**more common than you might think**).

Rigorous statistical inferences can only be drawn from properly designed experiments where we allocate individuals to treatments. Less powerful than these are "natural experiments," where a range of treatments have occurred by chance, and we can collect and analyse the data. Analysing natural patterns provide weak inferences, as they may have hidden biases. So, where you have a choice, you should try to carry out a properly controlled experiment rather than a correlational study that cannot infer causation.

## 11.3   General Principles of Experimental Design and Execution

In Chapter 4 we covered the importance of randomisation, replication, and blocking in designing an experiment in some detail. We will now see how these topics fit into the broader considerations involved in designing your project. The following list should be thought of as a checklist before you start data collection. If you follow this list, you will save yourself from experiencing difficulties later on.

### 11.3.1   What Are the Aims and Objectives of the Experiment?

The general aims (purpose) should be clear, and the objectives should be lucid and specific and expressed as: questions to be answered (how does pH affect reaction time?); hypotheses to be tested (Null hypothesis of no response); and effects to be estimated (for example, the mean increase in temperature is 5°C with a 95% confidence interval of between 4 and 6°C).

If you have both a primary and a secondary objective, you should make sure that the design of the experiment is effective and efficient for the primary objective and, ideally, also for the secondary objective if this can be achieved without compromising the primary objective. Think carefully about what you will be measuring or counting. Some variables will be relatively easy to observe — for example, the number of live seedlings in a pot. Others may be more difficult, so decide in advance which are of most interest to you. This way you can prioritise if there is not enough time for everything. Consider whether you will analyse each variable separately or whether you will combine any of them before analysis. For example, you might decide to multiply leaf width by leaf length to obtain an approximate estimate of leaf area.

### 11.3.2   What Population Are You Studying?

Work out the population about which you intend to generalise. Is it the individual, group, subspecies, or species? You need to check that the constraints within which you are working and ensure that you can generalise to this population. Make sure that the individuals in your study are representative of this population. John Harper (Harper, 1982) has pointed out the "trilemma" experienced by an experimental scientist, who may seek:

*Precision.* Can be achieved by using unique genotypes and carrying out the work in controlled environments. This should provide repeatable estimates with narrow confidence intervals but with perhaps less relevance to the "real world"?

*Realism.* May be achieved by studying individual plants in the field. This leads to low precision. Estimates will have very large confidence intervals because of the large amounts of random variation. Only large differences between populations will be detected as statistically significant.

*Generality.* It may be desirable to have information about treatment effects on a wide range of, for example, different soil types.

For each example, think of the population about which you can generalise. In the case of precision, you may end up being able to explain much of the variance in your system, but you are only able to generalise about the population of your greenhouse. At the other extreme, you might be able to generalise about an entire species range or ecosystem but not have much understanding to generalise. With limited resources, there is a danger of sacrificing both precision and realism and ending up with a shallow study, so be careful and get advice.

### 11.3.3  What Are the Experimental Units, and How Are They Grouped?

Experimental units (test tubes, plots, pots, plants, or animals) should not be able to influence each other and should be of a practical size and shape for the study. For example, if you are interested in the weight gain made by insects feeding on different species of plants, the experimental unit is the plant, but if you put several plants in one big pot, then they clearly influence each other, and you should take a mean value of what you have measured for each pot.

Pseudoreplication is very common in experiments, which is where you think you have more replicates than you do because some are not independent. This is often solvable but is something you should avoid at the experimental design stage, rather than ending up with a weaker experiment than you planned.

If you make the same measurement on each plant on several occasions (for example, measuring height), these will not be independent observations; they are repeated measures. A simple approach is to subtract, say, the first height from the last height and to analyse the increase in height over time. It is always sensible to obtain some estimate of the size of individuals before they have been exposed to any treatments, as this will probably explain some of the variance. This may well increase the precision of your experiment. You should ask advice about how to make use of such information.

### 11.3.4  What Are the Experimental Treatments, and What Is Your Control?

A control is a treatment to which nothing, or nothing of interest, is done. It is included in an experiment to show how the other treatments vary with respect to the background variation and how they vary with respect to each other. Is there a naturally defined control? If so, it should be included. For example, in estimating the effects of different levels of a chemical on an organism, it is wise to include the standard recommended amounts as well as applying none (the control).

### 11.3.5 Will Randomisation Be across All the Experimental Units or Constrained within Groups?

You allocate treatments to material at random to avoid bias and to be able to generalise the results. How to achieve this was covered in Chapter 4, both with and without blocks. A plan of the treatment allocations should be made. This is helpful in making records and may also be used to help explain any unexpected environmental variation that affects the results.

### 11.3.6 How Many Replicates Should You Have?

This is a common question, and you will need to talk to your supervisor and/or a statistician over this. You can calculate it or make an informed guess.

To calculate the number of replicates (called a **power** analysis), you need an estimate of the variability of the measurements (the standard error or standard deviation from a previous or pilot study) and the minimum size of the effect you are interested in. For example, if you are interested in population size in a seabird colony over a 10-year period, you might be interested in changes in mortality as small as 5%. You would need to design your experiment with sufficient power to detect this. If it is possible to carry out a power analysis, do it. It shows that you have planned the experiment as well as possible even if you do not subsequently detect anything. You may need to ask a statistician for advice about this because the required calculation depends on the design of the proposed experiment. Calculating the number of replicates promotes efficiency. If you have too few replicates, there will be no hope of detecting your required difference. Alternatively, if you have too many replicates, you may be wasting valuable time and money. Your project will, in all probability, have a small budget, and you must remain within it. Once the number of replicates has been fixed, it is important to remember to assess and record them separately.

If you do not have an estimate of the variability, and you do not know what size of effect might be important, then you are carrying out the experiment blind, so take some general precautions. Work out how many individuals in total you have the time and money to measure, and divide these between your treatments. If you have too few replicates in each treatment, you may well need to reduce the number of treatments and to test a simpler hypothesis.

### 11.3.7 What Resources (and Constraints) Do You Have?

You might have financial constraints that limit the number of replicates you can have. Another important factor is time. If it takes you half an hour to make observations on a badger set or to interview a subject, then there is a clear limit to the number of observations you can make. It is very important to be realistic here, so ask for advice or have a trial run. This is also important for judging how to balance this study with other commitments and meeting the final deadline.

### 11.3.8    How Will You Analyse Your Data?

You should define the specific questions that interest you as particular comparisons before carrying out the experiment. (For example, does treatment 3 have the same effect as treatment 4?) Check that you understand the proposed method of analysis and its constraints and that you will be able to carry it out using whatever sources you have available, whether this is a calculator, MINITAB on a personal computer, or even perhaps a more advanced package like Genstat, SAS, or SPSS. Getting familiar with your computer package will take longer than you think but will prove very satisfying in the end. If it will take some time before your experiment will produce results, practice the method of analysis using invented data. This experience will enable you to gain confidence.

### 11.3.9    Recording Data

In addition to experimental design, think about the practicalities. **Back up your data regularly.** Design a record sheet that is clear and easy to use, both when recording and when putting data into the computer. Each experimental unit should have a unique number (an ID), which might need to incorporate the identity of the operator if you are part of a larger study (for example, if several people are tagging trees, put your initials before your number). Ideally, the ID number should be used when recording, so that you are unaware of the treatment and can avoid prejudicing your results. Make sure you know exactly what you are going to record before you start and practice taking some observations before you start. Make notes as you go along of anything odd or that could be improved, and check for gross errors.

If your work is field based, one of the commonest things to go wrong is that your notebook gets wet. Waterproof notebooks are available from most outdoor shops. Pencils can still write in the wet and don't run. In addition to this, a digital camera is an excellent idea for taking illustrative photos and even backing up your notebook as images.

Remember, what can go wrong will go wrong. You should have a minimum of two copies of your data, both the electronic and hard copies, with regular backup using a data stick. Further to this, make comprehensive notes of everything that happens as you go along. It is infuriating trying to puzzle out exactly what you did, and when, some months after the event.

### 11.3.10    Other Considerations: Health and Safety, Ethical and Legal Issues

Proposed experiments in biomedical research must be approved by an independent committee of responsible people. For example, the allocation of treatments at random to people who are suffering from a disease may not be considered to be ethical if there is already evidence that one of the treatments is likely to be more efficacious than another. There are also strict controls on research work involving animals that your supervisor should guide you through and check you are following. However, there are also more general ethical considerations. It is unethical to waste resources, and important to minimise disturbance to animals or the environment, and you need to be aware of these issues.

Health and safety considerations are important to bear in mind to reduce accidents. There will also be a legal obligation on the part of your institution and yourself. Furthermore, marks are allocated for this in your project, so it is silly to ignore it or leave it until the last minute. Your supervisor should know the appropriate procedure, so make sure you discuss it. Student projects must comply with the law, relevant codes of practice, university local rules, and existing departmental safety procedures. Before you start work, whether in the laboratory or field, you may need training, vaccinations, or a health screening. The vaccinations, training, and protective equipment you will need to have or familiarise yourself with differs considerably for fieldwork in Antarctica and working with pathogens in a laboratory. Together with your supervisor, you have a legal obligation to draw up a list of risks, precautions taken to reduce those risks, and training and checkups that are needed. Ensure that you understand the job you will do, the risks involved, and how to mitigate them. If you are in any doubt, seek advice from your institution's safety officer.

### 11.3.10.1   Fieldwork

As part of fieldwork, you may need to be accompanied, or obtain permissions to enter and work on private land (this includes nature reserves). Plants must not be dug up and removed from a site. In advance, you will have to sort out protective clothing, a detailed map, compass (which you know how to use!), whistle, first-aid kit, and sufficient food. You will also need to have established contact protocols in case of emergency or something unforeseen in the fieldwork. You should always let somebody else know when and where you are going, and when you expect to be back. Contacts are especially important for overseas fieldwork or expeditions, as the ability to get advice and adapt may mean the difference between success and failure. If you are remote- or lone-working, you should carry a radio or mobile (check if there is reception!) with other researchers' phone numbers in it.

### 11.3.10.2   Labwork

If you are to work in a laboratory, you should already have sufficient experience in this area or have access to appropriate training before starting. You should meet the laboratory manager with your supervisor and discuss the proposed work. If you plan to use any chemicals that might pose a hazard to health, you should complete, in conjunction with your supervisor, the required form to outline the procedures that will be followed. The idea is that any possible hazards in your proposed work should be identified. There should also be a procedure in place for reporting potential problems, accidents, and near-misses.

## 11.4   Analysing Your Data and Writing the Report

The second half of this chapter provides some guidelines on how to analyse
the data and present your conclusions in a well-ordered scientific way. Con-
ventions differ between fields, so it is important to get some advice relevant
to your field or particular styles. It is worth having a look at a relevant journal
and following the "advice to authors" on its Website. When you are reading
around your study, you may find that there are standardised analyses for
certain types of data or styles that will make your data comparable with others.

### 11.4.1   Are Your Data Correct?

You will have made mistakes recording the data and transferring it from
your notebook into a spreadsheet, so it is essential to carry out a systematic
check. Are all the data present, correctly labelled, and in sequence? Are any
codes you have used in the correct range? For example, if you have 5 treatments coded 1 to 5, there should not be a 6, and if there are 6 replicates of
each treatment, there should not be 5 codes of "1" and 7 codes of "2."

When you identify a mistake, see if you can go back and correct it. If you
cannot work out what the value was, then you will have to exclude that data
point. Where data are missing, check that you have given them the right
code. Do not put a zero where you do not have data or it will alter your
group means. MINITAB uses a *, whereas SPSS uses a blank, so check that
it is the appropriate code for your statistical package.

Obvious errors can be spotted as outliers in box plots of each variable. Plots
of repeated observations over time can also reveal potential errors. Compare
the data from your original against a printout from the computer file. This will
identify any transcription errors. Whenever you have edited your data, consider
whether you wish to overwrite the previous, incorrect file of data or to save it
as a separate file. Two months into your project it is difficult to remember what
is in each file, so work out a clear file-naming system. It is important to keep
a note of what each file contains and to what extent the data have been checked.

There are numerous books written on how to write up a project. It is worth
having one on hand, but to avoid procrastination, only look up relevant sections
at a time. An excellent example is: Day and Gastel (2006) Writers' block is very
common. The only way around it is to work back from the final deadline and
set down a schedule of work to achieve it. This should include some spare time
to allow for emergencies. The key is to get a draft on paper that can serve as a
basis for revisions. If necessary, write your first draft as a list of bullet points
before rewriting it as full text. A large notebook in which you jot down dates,
thoughts, references, methods, and results as the study proceeds is invaluable.

Computer packages such as MINITAB enable you to store, edit, and analyse your data. You can export summary output and graphs from MINITAB
into a word processing package like Word (see Appendix A).

## 11.5 Structure

A report should have a clear structure, and in most cases you should start with the basic structure of a scientific paper. Make a list of subject headings and work on one section at a time. Many scientists write out all the section headings before even starting on any content. While you are working on other sections, you can put notes of any good ideas under the section heading.

Before you start, ask yourself: Who is my audience? What level of understanding do they have? What concepts do you need to explain rather than just to name? This is one of the most important parts of report writing and should be borne in mind throughout.

### 11.5.1 Title

This should be clear and not too verbose. It should include the subject or perhaps a statement of the results, e.g., "A Structure for Deoxyribose Nucleic Acid."

### 11.5.2 Abstract

This is the "take-home message," and possibly the only one that people will remember. It is a summary of the whole study in a paragraph or two. In this, you introduce the reader to what you have done, the findings, and the key points from the discussion. This is predominantly for carrying out searches of published material to determine whether a paper is relevant enough to continue reading. As students, you are less likely to care about someone searching for your work in the future, so the second reason is that you want to make the reader aware of the key points in advance so that they pay attention to the important parts when they get to them. The abstract is one of the hardest sections to write and is usually done last.

### 11.5.3 Introduction

This is the background that puts your study in context and contains a review of the relevant literature. You need to make the reader aware of what the current gap in knowledge is, so that it is obvious why you have carried out the study. The clearer this section is, the easier your discussion becomes.

### 11.5.4 Materials and Methods

This section contains a description of your protocol in sufficient detail that someone could replicate it. It should include details of

- Any pilot study
- The experimental or survey design

- The study populations
- The treatments (including any control)
- The reason for choosing the number of replicates
- The method of randomisation
- The methods of accounting for other sources of variation
- The variables measured, including units of measurement
- The criteria for assessing the outcome of the study
- Any possible sources of bias
- Any errors in execution (for instance, missing observations)
- Where and when the study took place
- A description of the apparatus and of any analytical methods used (or a reference to where these are fully described)
- Statistical analyses (or a reference to where these are fully described)

You may want to put details of the statistical method in a separate section as "Analysis." You should also state the statistical package you used (if any) and be aware of any standard scientific format or units. The idea is to enable someone else to be able to repeat your work or to take elements of it and modify them. *Never* gloss over mistakes, as this just makes you look vague.

### 11.5.5   Results

Be very clear about the difference between the results section and the discussion. The results sets out what you have collected and what it shows, in very specific terms. The discussion interprets these results and places them in the context of the wider literature.

So, present the data obtained with a brief explanation of the major trends revealed. You should illustrate the results section with tables and graphs (Section 11.7) and with reference to statistical analyses as appropriate. Ideally, you should provide exact "p-values" rather than simply "$p < 0.05$." It is important to concern yourself with your original objectives. In general, try to stick to the analysis you intended when you designed the experiment, rather than try to get every comparison out of the data.

### 11.5.6   Discussion

How do you interpret these results? How do the results affect our existing knowledge of the subject? Do they agree with current knowledge or theory? If so, in what way? Try to disprove this in your argument by looking for what other information would suggest that you should reject this conclusion. You should not be afraid to list the shortcomings of your own work and suggest further experiments that might resolve these problems. Trying to undermine yourself sounds counterproductive, but it is part of the scientific process, and you will look far less stupid if it is you yourself being thorough in this process rather than someone else. Many of the points you raise as criticisms would not have been obvious when you designed the experiment, so this is an important part of learning from your study.

Avoid repeating the results and analysis in the Discussion. You can be speculative; if your results do not support your hypothesis, is there an obvious reason why? Are you ready to reject your theory, or do you want to do another study incorporating the findings from this study?

It is important not to attempt to extrapolate the results of a subpopulation that you have studied to the whole population. For example, what you have discovered about the behaviour of worms on the local farm may well not be true for worms of a different species, or on a different soil type, or 200 km further north.

### 11.5.7 References

This is a list of the material cited in the report. You should quote sources of reference from the published literature or personal communication in the text, where appropriate, and list them fully in this section. You may be instructed to follow a particular "house style." This may seem picky to you, but it is essential to give accurate references and to double-check that the list contains all the required references and no others. If you use a reference software package like Endnote, it is much easier to keep track of and modify citations.

### 11.5.8 Acknowledgments

You should thank all those people or organizations that have helped you: your supervisor, members of the statistical and computing staff, the laboratory supervisor, anyone else who gave you access to data or resources, such as an important piece of equipment. Acknowledgments often get sentimental, especially when it is the last thing that people write before handing in a dissertation. In reality, it is a mixture of formal and personal; you need to recognize anyone else's intellectual or financial input or you are being negligent, but you might also want to add a personal note. For example:

> This work was funded by the BBSRC and BAS. I would like to thank Elaine Fitzcharles and Melody Clark for technical advice on sequencing. David Pearce helped with bacterial transformations and purification. Kate Cresswell and Mike Dunn bought me tea. Thanks.

### 11.5.9 Appendices

These are optional and contain items that are not required to follow the flow of the argument but may be useful to people who might follow up your work. Examples include raw data, calibration curves, species lists, and mathematical equations. It can be useful to provide the raw data but not necessary. If you have decided to use nonparametric statistics, it might be useful to suggest why your data are inappropriate for parametric analysis and show supporting graphs. In this book we have put statistical tables in appendices so that they do not disrupt the flow of the chapters.

## 11.6   The First Draft

When the above outline is in place, a first draft can appear. At this point it becomes obvious that there are gaps you need to fill or that some points should be made in a different order to emphasise connections. These points are attended to in preparing a second draft that should then be given to a friend whose opinion you value and (if it is the custom) to your supervisor for comments. It is worth getting two people to look over this: one who understands the subject and one who knows nothing. Usually, the person who knows nothing about the subject offers very valuable advice. More than the technical advisor, they will pick up on arguments that do not follow and points that need clarifying. These comments will vastly improve the clarity of the third draft. Do not be surprised if the reviewer covers it with comments in red ink. We all have to endure this because it is easier for others to spot our errors and inconsistencies than it is for us to see them ourselves. "Don't get mad — get even." Offer to reciprocate the task with another student, and then you can both vent your frustration.

Your report must be legible and well-written. Scientific reports are written in the past tense rather than in the present tense and use "we" rather than "I." It is traditional to use the passive rather than the active voice, although many journals are now starting to encourage the latter. So, traditionally, you would write: "The vegetation was sampled…," but if you wanted to write in the active: "we sampled the vegetation using five randomly positioned quadrats in each plot," that would be acceptable. Generic names of organisms take an upper case initial, and specific and subspecific names commonly have a lower case initial, and both are always in italics (or underlined): *Beta maritima*.

To be complete, you should follow the binomial name by its authority (an abbreviation for the name of the person credited with describing the species first): *Lolium perenne* L. (L. stands for Linnaeus). You should check spelling with a standard list and state which list you have used in the materials and methods section.

## 11.7   Illustrating Results

The point of illustrating results is to clarify difficult points and to provide a visual summary. People tend to remember graphs better than tables and understand concepts from them much quicker than they do with tables. In both cases, the caption should contain enough information to enable a reader to interpret the table or graph, but your interpretation of them should be in the discussion.

### 11.7.1    Graphs

Graphs are excellent for communicating qualitative aspects of data like shapes or orders of magnitude. For example, they can help you to show why you chose a nonlinear model over a straight-line model. It is important to label the axes of the graph, giving the units of measurement, and to present a suitable legend (caption). Give the graph a clear number, which you refer to in the text. With graphs, the number and legend is conventionally under the figure (Figure 11.1).

When you present a graph, you should also include a measure of the variability of the data. Most commonly, this is presented as a vertical bar labelled "SE mean" (standard error of a treatment mean after ANOVA, Chapter 7). Alternatively, you may put 95% confidence intervals around each mean. It is important to make clear which of these two possible measures of variability you have used.

### 11.7.2    Tables

Tables are required to convey quantitative features. A good table should display patterns and exceptions at a glance, but it will usually be necessary

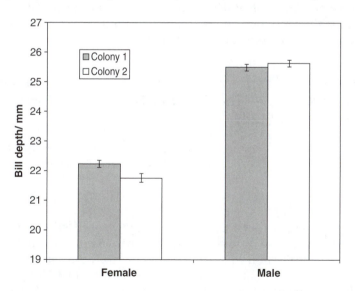

**FIGURE 11.1**
Mean male and female bill depths for two macaroni penguin colonies (*Eudyptes chrysolophus*). Bars show the standard error of the mean (unpublished data).

to comment in the text on the important points it makes. If tables are big and complex, they should go in an appendix to keep the argument clear.

As opposed to graphs and other figures, the convention with tables is to have the legend above the table. Some useful further guidelines come from Ehrenberg (1977):

- Standardise the number of decimal places, and where p-values are in the table.
- Show significance levels by starring significant results.
- It is easier to compare numbers in a column rather than in a row. The eye finds it easier to make comparisons vertically rather than horizontally.
- Ordering columns and rows by the size of the values in each cell helps to show pattern.

## 11.8   What It Is All About: Getting Through Your Project

We hope this book will be informative while you are starting out and a useful reference throughout your education or career. As with so many things in science, the real interest is in the relevance to real-life questions. Like turtles ready for the sea, no doubt you are itching to get out there and carry out your own study, so to help the transition from sand to sea, we have provided an example of experimental design and statistics in context. You might like to use it as a guide, but also criticise it and think about how you would improve on this if it were your own study.

## Project Report — The Effect of Fertiliser on Forage Rape: A Pilot Study

**A. N. Other**
*Small College, Largetown, Midshire*

### Abstract

To determine whether a larger-scale study was needed, we investigated the response of forage rape seedlings (*Brassica napus* L. ssp. *oleifera*) to fertilisation when grown in a putatively nutrient-poor soil. It was concluded that, for this soil, the addition of 67.5 gm$^2$ of a N:P:K:7:7:7 compound fertiliser would be adequate and that the effect of lower levels should be investigated in a subsequent pot experiment before carrying out a field experiment.

## Introduction

Forage rape (*Brassica napus* ssp. *oleifera*) is a crop grown for feeding to livestock in autumn and winter. An average fertiliser requirement for the crop is 125 kg ha$^{-1}$ N, 125 kg ha$^{-1}$P, 125 kg ha$^{-1}$ K$_2$0 (Lockhart and Wiseman, 1978). We wished to investigate the optimal amount of a compound fertiliser containing these nutrients to apply to seedlings growing on what was reputed to be a very nutrient-poor soil. A pilot study was carried out to check the suitability of the proposed experimental design and the range of amounts of fertiliser that should be used in a more detailed pot experiment before carrying out a field experiment. [The introduction should contain more detailed references to the literature to "set the scene." You should carry out a full literature search using key words such as "Growmore," "*Brassica napus*," "forage rape," and "fertiliser." You should make sure that you can do this for your field of interest.]

## Materials and Methods

We wished to generalise our results to the soil type (nutrient-poor Largetown loam), but for this pilot study we only had access to 2 ha of an arable field and to glasshouse space for 80 one-liter plant pots. We therefore chose 4 treatments: 67.5, 101, 135, and 270 g m$^{-2}$ of Growmore fertiliser. There were 20 replicate pots per treatment.

Results from this design were simulated in MINITAB to show that such experimental data could be analysed using ANOVA. If data from one or more pots were lost, the design could be analysed using a general linear model.

On September 10, 2006, a 1-cm layer of gravel was put into each of 80 pots. Each pot was given a unique number from 1 to 80 that was recorded on a spreadsheet to which details of treatment were added. Each pot was randomly assigned to a treatment and labelled. These labels were removed when all the pots had been filled and the seeds sown. This ensured that researchers were blind to the treatment during the growth and measurement stages. The Growmore granular fertiliser (N:P:K: 7:7:7) used in this study was ground to a powder and mixed throughout the soil. It was noted that the recommended rate for an "ordinary garden soil" is 135g m$^{-2}$. Pots were half filled with soil, the appropriate amount of fertiliser added, and the pots topped up with soil.

Ten seeds of *Brassica napus* were sown in each pot. The pots were arranged in the glasshouse in a wholly randomised design. The pots were examined daily and watered using rainwater from a common butt as required. The plants were thinned to 6 per pot on September 16 and to 4 per pot on September 20.

At harvest on October 15, the number of plants per pot was counted and their health scored (0 = perfect to 9 = all dead).

Shoot material from each pot was put into a labelled paper bag and oven-dried at 80°C for 48 h before being weighed.

This report concentrates on the effect of fertiliser on shoot dry-matter yield and plant health scores. The data were analysed using ANOVA, linear regression, and the Kruskal–Wallace test in MINITAB Release 14 (Ryan et al., 2005).

## Results

By September 16, germination had occurred in all pots. It was observed that, with increasing fertiliser, there were fewer plants, and these were of smaller size and were more yellow. There was very strong evidence that fertiliser affected shoot dry-matter yield ($F_{3,76} = 29.61$, $p < 0.001$). The mean and standard error is shown in Table 1 below.

**TABLE 1**

Summary Statistics for Shoot Dry Weight by Treatments (n = 20 for each)

| Fertiliser Treatment (kg/ha) | Mean (Shoot Dry Weight per Pot)(g) | SE Mean (g) |
|---|---|---|
| 67.5 | 0.48 | 0.036 |
| 101.0 | 0.36 | 0.020 |
| 135.0 | 0.32 | 0.026 |
| 270.0 | 0.15 | 0.011 |

It was noted that the mean shoot dry weight appeared to decline with increasing fertiliser. This was contrary to the expected out-come, and the decrease appeared to be linear. A linear regression analysis was carried out. Again, fertiliser was highly significant ($F_{1,78} = 83.12$, $p < 0.001$, $R^2$ (adj) = 0.51%). Thus, there is very strong evidence of a linear relationship. The fitted line equation is: *dry weight* = 0.5433 − 0.001494 *treatment*. This is shown in Figure 1.

The health scores per treatment are shown in Figure 2. To determine whether the fertiliser significantly affected the health score of the plants, a Kruskal–Wallis test was carried out. This was highly significant (H = 17.32, d.f. = 3, p = 0.001, adjusted for ties). The treatments differed significantly in the resultant health scores, with treatments receiving the highest amounts of fertiliser showing the poorest health (Figure 2).

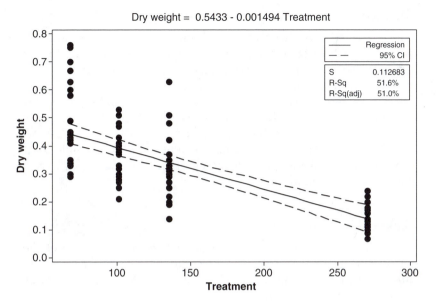

**FIGURE 1**

Fitted line plot to show the effect of fertiliser (kg ha⁻¹) on the shoot dry weight of forage rape g/pot. Dotted lines show the 95% confidence interval for the fitted line.

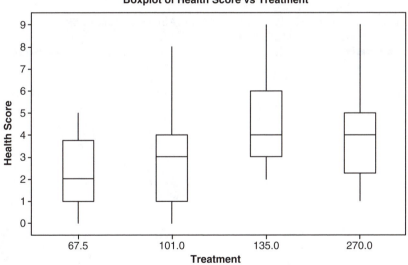

**FIGURE 2**

Box plot to show health score of plants for each fertiliser treatment (kg ha⁻¹).

## Discussion

The soil tested was of a surprisingly high nutrient status as indicated by the performance of the plants in the pots receiving the lowest amount of fertiliser that gave the highest yield. Greater amounts had a toxic effect. Perhaps an addition of even less than 67.5 g m$^{-2}$ (the lowest level of fertiliser) would give an even higher yield. It is hypothesized that applying no fertiliser would not produce the greatest yield, so in a further pot experiment, the levels of fertiliser should cover the range 0–101 g m$^{-2}$ to identify the optimum level. It is possible that plants grown for longer than 1 month might have a greater fertiliser requirement.

## Acknowledgments

I would like to thank the glasshouse staff for maintenance of the plants and the first-year students on Course 199 for harvesting them and recording the data, with help from P. Q. Smith. I thank E. D. Jones for computing advice.

## References

Lockhart, J. A. R. and Wiseman, A. J. L., *Introduction to Crop Husbandry*, 4th ed., Pergamon Press, Oxford, 1978.

Ryan, B. F., Joiner, B. L., and Cryer, J. D., *Minitab Handbook*, 5th ed., Duxbury Press, Belmont, CA, 2005.

# Appendix A

## *An Introduction to MINITAB*

### A.1  Conventions Used in This Book

We have developed a notation to describe how you do things using the MINITAB interface.

Each MINITAB instruction will usually be given like this (this is just an example):

```
Stat>Basic Statistics>Display Descriptive Statistics: Variables
Worms
```

It means: *"Select 'Display Descriptive Statistics' submenu of the 'Basic Statistics' submenu of the 'Stat' main menu item. Select the variable 'Worms' for the 'Variables' box."*

- **Bold text** is what you *see on the screen* as labels on dialog boxes or radio buttons.
- **Normal type** represents *your input*, i.e., what you put in the dialog boxes.

Where the command opens up subwindows, and where these have tabs as well as boxes, we have tried to adopt a similar approach. e.g.,

```
Graph>Histogram Simple Graph Variable worms DataView>DataDisplay
Bars Area
```

This invites you to

*"Select 'Histogram' submenu from 'Graph' Main menu item and choose 'Simple' from the gallery. Put the 'worms' column name in the 'Graph' variables box. Select 'Data View' button, select the 'Data Display' tab and check the 'Bars' and 'Area' boxes."*

## A.2  Starting Up

When you first start the MINITAB Application, you see two windows on the screen. The top one is the Session window, in which the output from the program's activities appears. Below is a Data window consisting of columns and rows. Hidden from view, but accessible by **Window>Project Manager,** is the Project Manager, a control centre for the whole program. It is a Management Console or Explorer-like window with various selectable objects in the left panel, and a right panel to display information about whatever has been selected.

## A.3  Help

The most essential menu item is the online **Help**. It has a number of sub-menus, the most useful of which are

- **Help**. A complete manual for the program.
- **Tutorials.** A series of guided sessions to introduce the capabilities of the program. The first of these is largely accessible without any previous knowledge of statistics, but sessions 2 to 5 assume that you already know about statistics and want to learn about the program.
- **StatGuide**. An interactive statistical reference manual.

## A.4  Data Entry

Entering data is similar to any spreadsheet application. The Data window looks just like a conventional spreadsheet. You can put a name in the top row, under where it says C1, C2 … and enter the data down the column. Columns can be referred to by their names, or as "C1, C2 etc." Editing operations on whole columns are found under the **Editor** object on the main menu bar, and operations affecting Cells are under **Edit.** Data can be numbers or letters, but if any cell in a column contains letters, the whole column is treated as letters, including cells containing only numerals. If you accidentally cause this conversion, you can change it back again (after removing the offending characters) using **Data>ChangeDataType>Text to Numeric**.

## A.5  Looking at the Worms Data

We briefly look at the operations needed to study the worms data set of Chapter 1. We assume you have entered the data into a column with the heading "length."

### A.5.1  Display the Information in a Column

This is not all that useful unless you want to put the data into a report — you can see the data in the Data Window, but if you want to print the contents of a column in the Session window, then use **Data>Display Data Columns ... to display** length

This is what you see in the Session window:

```
MTB > Print 'length'.
```

**Data Display**

```
Length
   6.0   8.0   9.0   9.5   9.5  10.0   10.5  11.5  12.5  13.5
```

### A.5.2  Dot Plot the Data

Using **Graph>Dot Plot: Simple Graph variables** length, you produce Figure 1.1a.

### A.5.3  Produce Summary Statistics

Use **Stat>Basic Statistics>Display Descriptive Statistics Variables** length

```
MTB > Describe 'length'
```

**Descriptive Statistics: Length**

```
Variable    N   N*     Mean  SE Mean StDev Minimum      Q1  Median
Length     10    0   10.000    0.687 2.173   6.000   8.750   9.750

Variable        Q3  Maximum
Length      11.750   13.500
```

### A.5.4  Box Plot

The box-and-whisker plot is a very useful exploratory tool that conveys information about the overall distribution of the data without representing

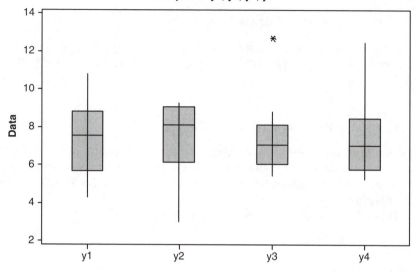

**FIGURE A.1**

Box plots of 4 data sets used in Chapter 8. The line in the middle of the box is the median, the limits of the box are the upper and lower quartiles. Note the outlier (the * symbol) in the plot for y3.

every point. Figure A.1 shows a multiple box plot for four columns of data named y1, y2, y3, y4. This was produced by using:

```
Graph>Boxplot Multiple Y's Graph variables y1 y2 y3 y4.
```

### A.5.5 Histogram

To produce a simple frequency histogram (Figure 1.1b) use **Graph>Histogram: Simple Graph variables** length. In fact you will not get exactly Figure 1.1b because it seems MINITAB is slightly arbitrary about whether a point on a class boundary goes into the class above or the one below. You force your own class boundaries like this

- Double-click the x-axis the displayed histogram
- Choose **Binning>Interval Type** cutpoint
- Choose your own cutpoints. Fill in the window like this (Figure A.2).

You can easily convert the y-axis to relative frequency density by double-clicking on the y-axis and choosing **Type>Scale Type** Density to produce Figure 1.5a.

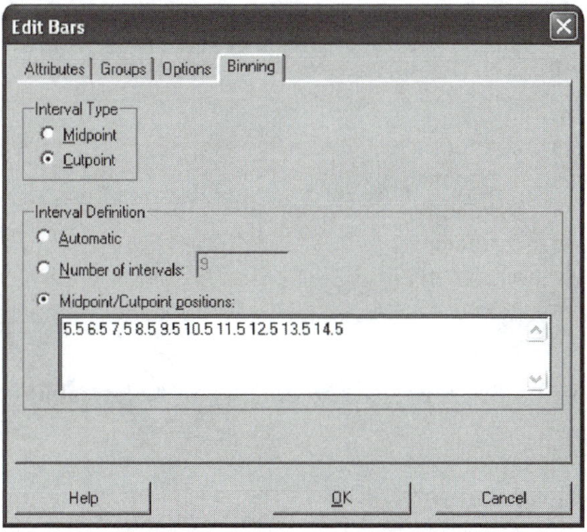

**FIGURE A.2**
Setting your own cutpoints in histogram.

## A.6 Updating Graphs

Unlike most spreadsheets, graphical output will normally not change when you make changes to the underlying data. You can make a graph update without redrawing the whole thing by right-clicking on the graph and selecting **Update Graph Now** (greyed out unless the data have, in fact, changed). You can make that particular graph update every time the data change by means of **Right Click>Update Graph Automatically.** Graphs are interactive in the sense that hovering the cursor over any feature gives appropriate information about it (e.g., its value, the row of the data involved, or the summary statistic that is being represented) as a "screen tip." Almost any aspect of a graph can be edited by right-clicking on it or, if you have trouble selecting the feature you want, using **Editor>Select Item**. Many of these features are demonstrated in **Help>Tutorials Session One**.

## A.7 Stacking and Unstacking — A Useful Trick

We quite often find we need to convert data from a series of columns representing measurements of a series of different groups to a single column with an indicator saying which group each measure comes from. In Chapter 3 (Section 3.3) we have data from pairs of twin sheep raised on different diets:

```
MTB > Print 'old' 'new'.
```

**Data Display**

```
Row      Old       New
  1  3.40000  4.50000
  2  3.90000  4.80000
  3  4.20000  5.70000
  4  4.50000  5.90000
  5  3.60000  4.30000
  6  2.90000  3.60000
  7  3.20000  4.20000
```

Later on, in Section 3.7, we need the data as an "edge table":

**Data Display**

```
Row  Sheepweigh  Diet
  1         3.4   old
  2         3.9   old
  3         4.2   old
  4         4.5   old
  5         3.6   old
  6         2.9   old
  7         3.2   old
  8         4.5   new
  9         4.8   new
 10         5.7   new
 11         5.9   new
 12         4.3   new
 13         3.6   new
```

To get from the first to the second form we use

> **Data>Stack>Columns Stack ... columns** old new **Store ...**
> **in Column of current worksheet** sheepweigh **Store subscripts in** diet
> **Use variable names ...**

To reverse the process use **Data>Unstack Columns Unstack the data in** sheepweigh **Using the subscripts in** diet **Store unstacked data: After last column ... Name the columns ...**

---

## A.8  Looking Up Probabilities

MINITAB effectively contains a complete set of statistical tables for almost every distribution you are likely to need. You can use this in three different ways.

### A.8.1 Probability Density

What is p(X) in a given distribution? This only applies to discrete distributions such as the Binomial or Poisson (see Chapter 1, Box 1.3). In Table 1.3 we ask what is the probability that a family of 3 children has no boys, given that the probability that any child is a boy is 0.5?

**Calc>Probability Distributions>Binomial Probability Number of trials** 3 **Probability of success** 0.5 **Input constant** 0

(Note that here you must remember to select the **Probability** option as it is not the default.)

```
MTB > PDF 0;
SUBC>    Binomial 3 0.5.
```

**Probability Density Function**

```
Binomial with n = 3 and p = 0.5
x   P(X = x)
0      0.125
```

You can also get the whole distribution by setting the possible values of X = number of boys into a column (we will call it "boys"). The command is similar to before:

**Calc>Probability Distributions>Binomial Probability Number of trials** 3 **Probability of success** 0.5 **Input column** boys

```
MTB > PDF 'boys';
SUBC>    Binomial 3 0.5.
```

**Probability Density Function**

```
Binomial with n = 3 and p = 0.5
x   P(X = x)
0      0.125
1      0.375
2      0.375
3      0.125
```

### A.8.2 Cumulative Density

This is a more common requirement. What is the probability in the Standard Normal distribution that $z \leq -2.00$? The answer, according to Table C.1, is 0.0228. MINITAB will find this value as

**Calc>Probability Distributions>Normal Cumulative Probability Mean** 0.0 **Standard Deviation** 1.0 **Input constant** −2.0

```
MTB > CDF −2.0;
SUBC> Normal 0.0 1.0.
```

**Cumulative Distribution Function**

```
Normal with mean = 0 and standard deviation = 1
  x     P(X ≤ x)
−2   0.0227501
```

It is important to remember that the Cumulative Distribution Function always gives the left tail of the distribution. Thus, if we replace –2.0 by +2.0 in the above example, we get this:

```
MTB > CDF 2.0;
SUBC> Normal 0.0 1.0.
```

**Cumulative Distribution Function**

```
Normal with mean = 0 and standard deviation = 1
x      P(X ≤ x)
2    0.977250
```

If we are interested in the right tail here (which usually we would be), we must calculate

$$P(X > 2.0) = 1 - P(X \leq 2.0) = 1 - 0.977250 = 0.0227500$$

### A.8.3 Inverse Cumulative Density

What is the value of chi-square that is exceeded only 5% of the time when there is one degree of freedom? Table C.6 says "3.84." Because the distribution functions work on the left tail, we must rephrase the question as "What is the value in the chi-squared distribution with 1 degree of freedom such that it is *not exceeded* 95% of the time?" The command is

**Calc>Probability Distributions Inverse Cumulative Probability Noncentrality … 0.0 Degrees of Freedom 1 Input constant 0.95**

```
MTB  > InvCDF 0.95;
SUBC > ChiSquare 1.
```

**Inverse Cumulative Distribution Function**

```
Chi−Square with 1 DF
P(X ≤ x)          x
    0.95   3.84146
```

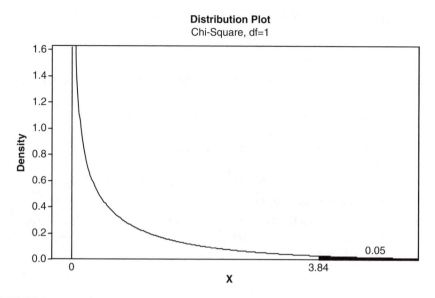

**FIGURE A.3**
Value of chi-square that is exceeded 5% of the time is 3.84 using Graph>Probability Distribution Plot.

### A.8.4 Graphical Output — New in MINITAB 15

The left- and right-tail issue causes much confusion, which is best dealt with by visualizing the output. In the latest release of MINITAB this is possible. Using the example of A8.3, we have

```
Graph>Probability Distribution Plot View Probability (OK)
Distribution Chi-Square Degrees of Freedom 1 Shaded Area Define
… by Probability Right Tail Probability 0.05.
```

The result is Figure A.3.

## A.9 Report Writer

MINITAB has a very useful feature to help you create reports on your activities.

Find the *Project Manager* window (use **Window>Project Manager**) and then click on the ReportPad item in the left window.

The right window is blank with the heading *Minitab Project Report*.

To add an item to the report:

- Place the cursor next to the Bold heading for it in the Session Window.

- Right click and select **Append Section to Report**.

To add a graph:

- Click anywhere on the graph and **Right Click>Append Graph to Report**.

You can also type anything you like in the way of explanatory text around the results.

To print the report, or save as a separate document,

- Place the cursor in the Report section of the ReportPad window.
- To print **File>Print Report**.
- To save **File>Save Report As.** The report is saved as Rich Text Format (.RTF) format, which can be read into most word processors for further editing).

## A.10 The MINITAB Command Line

Like most modern statistical packages, MINITAB has evolved from a program that was driven by sequences of commands typed in at a keyboard. This interface lives on out of sight but can be brought into view for the current session by making the Session window active (click anywhere in it) and selecting **Editor>Enable Commands.** Each time you use the point-and-click interface, you will see the corresponding commands appear in the session window. Once commands are enabled typing in

> **MTB >** describe 'length' (usual convention — **bold** is what you see, normal is what you type)

it will produce the same output as the **Stat>Display Descriptive Statistics...** in Section A.5.3.

We have left these commands visible in the examples in this Appendix, though they are not shown elsewhere. If you want to arrange so that commands are always enabled, use **Tools>Options Session Window>Submitting Commands** to set your preferences.

> **MTB >** help invokes the Help relevant to the command line interface.

Although it may seem like an exotic curiosity, the Command line interface has a number of uses:

- It represents a concise record of what you did to get a particular result. The whole of your session is stored in the form of Command line entries in the **History** object of the Project Manager.

- Using the Command Line Editor **Edit>Command Line Editor** you can repeat a complicated sequence of operations, making minor changes as you go.

- If you want to automate a process in MINITAB, the script is written as a series of command-line commands.

Although most major statistical packages have hidden their command interface in a similar way, there is a striking exception, which is the free statistical package R. Although its command language is different, there are sufficient similarities for us to suggest that learning MINITAB's command interface by seeing what commands are generated from the point-and-click interface is a good preparation for the difficulties of a purely command-driven environment.

## A.11 Saving Your Session

**File>Save Project** saves the entire session so that if you reopen it with **File>Open Project**, you find matters exactly as you left them. Other items on the File menu provide a number of useful facilities, including querying external databases and reading and writing Worksheets in a number of formats compatible with other software.

# Appendix B

## Statistical Power and Sample Size

The power of a statistical test is defined as the probability that the Null hypothesis will be rejected, given that it is indeed false. Intuitively, it is the capacity of a test to detect that the Null hypothesis is false (and therefore that the Research hypothesis is true). The importance of power in the design of experiments is that it can influence our choice of sample size, and increase our confidence in the design of an experiment.

Recall that if we reject a Research hypothesis when it is in fact true (equivalent to accepting $H_0$ when it is in fact false) we commit a Type II error; the probability that this will happen is usually given the symbol $\beta$; so it follows that the power of a test is the probability we will *not* commit a Type II error, or $1-\beta$.

Consider the case where we are conducting a single sample z-test (section 2.4) on a sample of data from a population for which we know the standard deviation $\sigma = 10$. The test is to be 1-tailed. Using the techniques introduced in section 2.4 we calculate

$$z = \frac{\bar{x} - \mu_0}{\sigma / \sqrt{n}}$$

The critical value for a one tailed z-test is 1.64 (looked up in Appendix C1, or using the computer), so we would reject $H_0$ and therefore accept $H_A$ if $z > 1.64$. There are two situations to consider:

1. $H_0$ is in fact the case. If so, on average $\bar{x} = \mu_0$ and so on average

$$z = \frac{\bar{x} - \mu_0}{\sigma / \sqrt{n}} = \frac{0}{\sigma / \sqrt{n}} = 0$$

2. $H_A$ is the case. If so, on average $\bar{x} = \mu_1$ and

$$z = \frac{\bar{x} - \mu_0}{\sigma / \sqrt{n}} = \frac{\mu_A - \mu_0}{\sigma / \sqrt{n}} ;$$

it is convenient to tidy this up to

$$z = \sqrt{n} \times \frac{\mu_A - \mu_0}{\sigma}.$$

The term

$$\frac{\mu_A - \mu_0}{\sigma}$$

is called the **effect size** and represents, in units of the population standard deviation, the magnitude of the effect we are looking for with our test.

These two situations are shown graphically in Figure B.1 for the case where our sample size $n$ is 100. The left (continuous) curve shows the expected distribution of $z$ given that the Null hypothesis is true. The right (dashed)

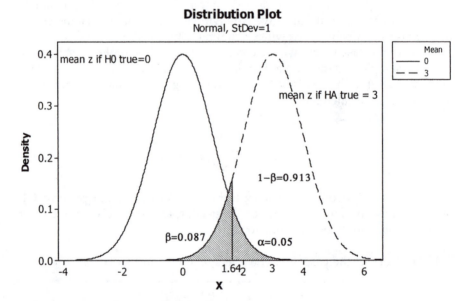

**FIGURE B.1**
Expected distribution of z in a single sample z-test if $H_0$ is true (left, solid, curve), or where a specific alternative hypothesis is true (right, dotted, curve). For a sample size of $n = 100$ and a population standard deviation of $\sigma = 10$ and a difference $\mu_A - \mu_0 = 3$ the dashed curve is centred on

$$z = \sqrt{n} \times \frac{\mu_A - \mu_0}{\sigma} = \sqrt{100} \times \frac{3}{10} = 10 \times \frac{3}{10} = 3.$$

The area to the right of 1.64 is approximately 0.05 in the left curve, so this is the 5% critical value (one tailed). The area to the left of this point in the right curve lies in the region where we would accept H0, so it gives us $\beta$, the probability we would reject $H_A$ when in fact it was true, ie the probability we would commit a Type II error if $H_A$ is true.

curve shows the case where the Research hypothesis is true for difference between the Null hypothesis and the research hypothesis of $\mu_A - \mu_0 = 3$. Thus the dashed curve is centred on

$$z = \sqrt{n} \times \frac{\mu_A - \mu_0}{\sigma} = \sqrt{100} \times \frac{3}{10} = 10 \times \frac{3}{10} = 3$$

(note that we deliberately chose the sample size of 100 to make the sums easy!) Because the distributions are plotted in terms of $z$ the standard deviations of the curves we see are both 1. We can now determine the power of our test by choosing the value on the x-axis corresponding to the case where we would reject $H_0$ and accept $H_A$ — we have already looked this up and agreed that it is approximately 1.64. The graph shows $\alpha$, the probability we will reject $H_0$ given that it is true, as 0.05, the hatched area in the continuous curve to the right of $z = 1.64$. The hatched area in the dashed curve, to the left of $z = 1.64$, is $\beta$, the probability we will reject $H_A$ given that it is in fact true. In the example this is 0.087 on the assumption that:

- $\sigma$ is 10
- $n$ is 100
- We have chosen $\alpha$ to be 0.05, and we have chosen to do a 1-tailed test

So the conclusion is that the test would have a statistical power of $1-\beta = 0.913$, or slightly more than 90%.

There are two problems with putting this understanding to practical use

1. The manual calculations are somewhat fiddly, especially when the test of interest is more complicated than a one sample z-test.
2. What would usually like to know is, given an effect size and a population standard deviation, what sample size would we need for a given power? Reversing the calculations to solve for $n$ is difficult and requires special tables (found for example in Samuels and Witmer, 2003).

Fortunately computers come to our aid. In MINITAB the command **Stat>Power and Sample Size>1-sample Z Sample sizes:** 100 **Differences** 3 **Standard deviation** 10 **Options>Alternative hypothesis Greater than Significance level** 0.05 generates the following:

```
Power and Sample Size
1-Sample Z Test

Testing mean = null (versus > null)
Calculating power for mean = null + difference
Alpha = 0.05   Assumed standard deviation = 10

               Sample
Difference      Size      Power
         3       100   0.912315
```

which confirms our manual result. Note that the Difference is given in the original scale of measurement. If you wanted to use effect size as defined above, put one or more values of effect size in the Differences box and set the standard deviation to 1.

To illustrate the use of this technique we will find sample sizes we need for different powers and effect sizes in the simulation of data in Section 4.5. We use **Stat>Power and Sample Size>2-sample t Sample sizes: 9 20 30 Power values .7 .8 .9 Standard deviation** 1 **Options>Alternative hypothesis Not equal Significance level** 0.05. Thus we are exploring sample sizes 9, 20, 30 (note that this is for each group) with Power values from 70% to 90%. Here is the output:

**Power and Sample Size**
2-Sample t Test

Testing mean 1 = mean 2 (versus not =)
Calculating power for mean 1 = mean 2 + difference
Alpha = 0.05  Assumed standard deviation = 1

Sample

| Size | Power | Difference |
|------|-------|------------|
| 9    | 0.7   | 1.24710    |
| 9    | 0.8   | 1.40693    |
| 9    | 0.9   | 1.62879    |
| 20   | 0.7   | 0.80613    |
| 20   | 0.8   | 0.90913    |
| 20   | 0.9   | 1.05199    |
| 30   | 0.7   | 0.65231    |
| 30   | 0.8   | 0.73562    |
| 30   | 0.9   | 0.85117    |

The sample size is for each group.

This shows that if we wanted 80% power (a reasonable compromise – one should not be too optimistic about the power of such a test) we could detect an effect size of 0.736 with groups of 30 individuals, but it would have to be more than 1.41 for us to detect it with groups of only 9, as we were proposing. As a rule of thumb, an effect size of 0.4 is considered "small" whereas one of 1.4 is considered "large," so in this case we might feel that groups of 20, or even 30, would be preferable. Clearly it is important for us to know how big an effect we might find in our experiment, which means having a plausible estimate of the likely population standard deviation, and an expectation as to how big the difference between the two groups would be if the research hypothesis were correct. Thus we may need some pilot data to give us an idea of the variability of our study material.

In the latest release of Minitab (15) you can also get a graphical representation of the situation (Figure B.2). This allows you to estimate the Power of a wider range of possibilities than just those tabulated in the printed output.

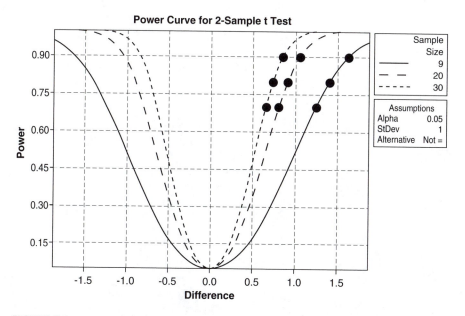

**FIGURE B.2**

Family of Power curves for a 2-sample t-test. The 3 curves are for different sample sizes. By choosing StDev=1 the differences are expressed in units of effect size

$$\frac{\mu_A - \mu_0}{\sigma}.$$

The rejection probability for $H_0$ is assumed to be 0.05, and the test is to be 2 tailed. The symbols are the values represented in the printed numerical output.

# Appendix C

# Statistical Tables

**TABLE C.1**

The Standardised Normal Distribution

The distribution tabulated is that of the normal distribution with mean 0 and standard deviation 1. For each value of the standardised normal deviate $z$, the proportion $P$ of the distribution less than zero is given. For a normal distribution with mean $\mu$ and variance $\sigma^2$, the proportion of the distribution less than some particular value $x$ is obtained by calculating $z = (x - \mu)/\sigma$ and reading the proportion corresponding to this value of $z$.

| $z$ | $P$ | $z$ | $P$ | $z$ | $P$ | $z$ | $P$ |
|---|---|---|---|---|---|---|---|
| −4.00 | 0.00003 | −1.50 | 0.0668 | 0.00 | 0.5000 | 1.55 | 0.9394 |
| −3.50 | 0.00023 | −1.45 | 0.0735 | 0.05 | 0.5199 | 1.60 | 0.9452 |
| −3.0 | 0.0014 | −1.40 | 0.0808 | 0.10 | 0.5398 | 1.65 | 0.9505 |
| −2.95 | 0.0016 | −1.35 | 0.0885 | 0.15 | 0.5596 | 1.70 | 0.9554 |
| −2.90 | 0.0019 | −1.30 | 0.0968 | 0.20 | 0.5793 | 1.75 | 0.9599 |
| −2.85 | 0.0022 | −1.25 | 0.1056 | 0.25 | 0.5987 | 1.80 | 0.9641 |
| −2.80 | 0.0026 | −1.20 | 0.1151 | 0.30 | 0.6179 | 1.85 | 0.9678 |
| −2.75 | 0.0030 | −1.15 | 0.1251 | 0.35 | 0.6368 | 1.90 | 0.9713 |
| −2.70 | 0.0035 | −1.10 | 0.1357 | 0.40 | 0.6554 | 1.95 | 0.9744 |
| −2.65 | 0.0040 | −1.05 | 0.1469 | 0.45 | 0.6736 | 2.00 | 0.9772 |
| −2.60 | 0.0047 | −1.00 | 0.1587 | 0.50 | 0.6915 | 2.05 | 0.9798 |
| −2.55 | 0.0054 | −0.95 | 0.1711 | 0.55 | 0.7088 | 2.10 | 0.9821 |
| −2.50 | 0.0062 | −0.90 | 0.1841 | 0.60 | 0.7257 | 2.15 | 0.9842 |
| −2.45 | 0.0071 | −0.85 | 0.1977 | 0.65 | 0.7422 | 2.20 | 0.9861 |
| −2.40 | 0.0082 | −0.80 | 0.2119 | 0.70 | 0.7580 | 2.25 | 0.9878 |
| −2.35 | 0.0094 | −0.75 | 0.2266 | 0.75 | 0.7734 | 2.30 | 0.9893 |
| −2.30 | 0.0107 | −0.70 | 0.2420 | 0.80 | 0.7881 | 2.35 | 0.9906 |
| −2.25 | 0.0122 | −0.65 | 0.2578 | 0.85 | 0.8023 | 2.40 | 0.9918 |
| −2.20 | 0.0139 | −0.60 | 0.2743 | 0.90 | 0.8159 | 2.45 | 0.9929 |
| −2.15 | 0.0158 | −0.55 | 0.2912 | 0.95 | 0.8289 | 2.50 | 0.9938 |
| −2.10 | 0.0179 | −0.50 | 0.3085 | 1.00 | 0.8413 | 2.55 | 0.9946 |
| −2.05 | 0.0202 | −0.45 | 0.3264 | 1.05 | 0.8531 | 2.60 | 0.9953 |
| −2.00 | 0.0228 | −0.40 | 0.3446 | 1.10 | 0.8643 | 2.65 | 0.9960 |
| −1.95 | 0.0256 | −0.35 | 0.3632 | 1.15 | 0.8749 | 2.70 | 0.9965 |
| −1.90 | 0.0287 | −0.30 | 0.3821 | 1.20 | 0.8849 | 2.75 | 0.9970 |
| −1.85 | 0.0322 | −0.25 | 0.4013 | 1.25 | 0.8944 | 2.80 | 0.9974 |
| −1.80 | 0.0359 | −0.20 | 0.4207 | 1.30 | 0.9032 | 2.85 | 0.9978 |
| −1.75 | 0.0401 | −0.15 | 0.4404 | 1.35 | 0.9115 | 2.90 | 0.9981 |
| −1.70 | 0.0446 | −0.10 | 0.4602 | 1.40 | 0.9192 | 2.95 | 0.9984 |
| −1.65 | 0.0495 | −0.05 | 0.4801 | 1.45 | 0.9265 | 3.00 | 0.9986 |
| −1.60 | 0.0548 | 0.00 | 0.5000 | 1.50 | 0.9332 | 3.50 | 0.99977 |
| −1.55 | 0.0606 | | | | | 4.00 | 0.99997 |

## TABLE C.2

The Student's *t*-Distribution

This table gives the value of *t* for which a particular percentage *P* of the Student's *t*-distribution lies outside the range –*t* to +*t*. These values of *t* are tabulated for various degrees of freedom.

| Degrees of freedom | P | | | | | | | |
|---|---|---|---|---|---|---|---|---|
| | 50 | 20 | 10 | 5 | 2 | 1 | 0.2 | 0.1 |
| 1 | 1.00 | 3.08 | 6.31 | 12.7 | 31.8 | 63.7 | 318 | 637 |
| 2 | 0.82 | 1.89 | 2.92 | 4.30 | 6.96 | 9.92 | 22.3 | 31.6 |
| 3 | 0.76 | 1.64 | 2.35 | 3.18 | 4.54 | 5.84 | 10.2 | 12.9 |
| 4 | 0.74 | 1.53 | 2.13 | 2.78 | 3.75 | 4.60 | 7.17 | 8.61 |
| 5 | 0.73 | 1.48 | 2.02 | 2.57 | 3.36 | 4.03 | 5.89 | 6.87 |
| 6 | 0.72 | 1.44 | 1.94 | 2.45 | 3.14 | 3.71 | 5.21 | 5.96 |
| 7 | 0.71 | 1.42 | 1.89 | 2.36 | 3.00 | 3.50 | 4.79 | 5.41 |
| 8 | 0.71 | 1.40 | 1.86 | 2.31 | 2.90 | 3.36 | 4.50 | 5.04 |
| 9 | 0.70 | 1.38 | 1.83 | 2.26 | 2.82 | 3.25 | 4.30 | 4.78 |
| 10 | 0.70 | 1.37 | 1.81 | 2.23 | 2.76 | 3.17 | 4.14 | 4.59 |
| 12 | 0.70 | 1.36 | 1.78 | 2.18 | 2.68 | 3.05 | 3.93 | 4.32 |
| 15 | 0.69 | 1.34 | 1.75 | 2.13 | 2.60 | 2.95 | 3.73 | 4.07 |
| 20 | 0.69 | 1.32 | 1.72 | 2.09 | 2.53 | 2.85 | 3.55 | 3.85 |
| 24 | 0.68 | 1.32 | 1.71 | 2.06 | 2.49 | 2.80 | 3.47 | 3.75 |
| 30 | 0.68 | 1.31 | 1.70 | 2.04 | 2.46 | 2.75 | 3.39 | 3.65 |
| 40 | 0.68 | 1.30 | 1.68 | 2.02 | 2.42 | 2.70 | 3.31 | 3.55 |
| 60 | 0.68 | 1.30 | 1.67 | 2.00 | 2.39 | 2.66 | 3.32 | 3.46 |
| ∞ | 0.67 | 1.28 | 1.64 | 1.96 | 2.33 | 2.58 | 3.09 | 3.29 |

**TABLE C.3**

The *F*-Distribution: 5% Points

These tables give the values of *F* for which a given percentage of the *F*-distribution is greater than *F*.

| $n_2$ | $n_1$ | | | | | | | | | | |
|---|---|---|---|---|---|---|---|---|---|---|---|
| | 1 | 2 | 3 | 4 | 5 | 6 | 7 | 8 | 10 | 12 | 24 |
| 2 | 18.5 | 19.0 | 19.2 | 19.2 | 19.3 | 19.3 | 19.4 | 19.4 | 19.4 | 19.4 | 19.5 |
| 3 | 10.1 | 9.55 | 9.28 | 9.12 | 9.01 | 8.94 | 8.89 | 8.85 | 8.79 | 8.74 | 8.64 |
| 4 | 7.71 | 6.94 | 6.59 | 6.39 | 6.26 | 6.16 | 6.09 | 6.04 | 5.96 | 5.91 | 5.77 |
| 5 | 6.61 | 5.79 | 5.41 | 5.19 | 5.05 | 4.95 | 4.88 | 4.82 | 4.74 | 4.68 | 4.53 |
| 6 | 5.99 | 5.14 | 4.76 | 4.53 | 4.39 | 4.28 | 4.21 | 4.15 | 4.06 | 4.00 | 3.84 |
| 7 | 5.59 | 4.74 | 4.35 | 4.12 | 3.97 | 3.87 | 3.79 | 3.73 | 3.64 | 3.57 | 3.41 |
| 8 | 5.32 | 4.46 | 4.07 | 3.84 | 3.69 | 3.58 | 3.50 | 3.44 | 3.35 | 3.28 | 3.12 |
| 9 | 5.12 | 4.26 | 3.86 | 3.63 | 3.48 | 3.37 | 3.29 | 3.23 | 3.14 | 3.07 | 2.90 |
| 10 | 4.96 | 4.10 | 3.71 | 3.48 | 3.33 | 3.22 | 3.14 | 3.07 | 2.98 | 2.91 | 2.74 |
| 12 | 4.75 | 3.89 | 3.49 | 3.26 | 3.11 | 3.00 | 2.91 | 2.85 | 2.75 | 2.69 | 2.51 |
| 15 | 4.54 | 3.68 | 3.29 | 3.06 | 2.90 | 2.79 | 2.71 | 2.64 | 2.54 | 2.48 | 2.29 |
| 20 | 4.35 | 3.49 | 3.10 | 2.87 | 2.71 | 2.60 | 2.51 | 2.45 | 2.35 | 2.28 | 2.08 |
| 24 | 4.26 | 3.40 | 3.01 | 2.78 | 2.62 | 2.51 | 2.42 | 2.36 | 2.25 | 2.18 | 1.98 |
| 30 | 4.17 | 3.32 | 2.92 | 2.69 | 2.53 | 2.42 | 2.33 | 2.27 | 2.16 | 2.09 | 1.89 |
| 40 | 4.08 | 3.23 | 2.84 | 2.61 | 2.45 | 2.34 | 2.25 | 2.18 | 2.08 | 2.00 | 1.79 |
| 60 | 4.00 | 3.15 | 2.76 | 2.53 | 2.37 | 2.25 | 2.17 | 2.10 | 1.99 | 1.92 | 1.70 |

**TABLE C.4**

The *F*-Distribution: 1% Points

| $n_2$ | $n_1$ | | | | | | | | | | |
|---|---|---|---|---|---|---|---|---|---|---|---|
| | 1 | 2 | 3 | 4 | 5 | 6 | 7 | 8 | 10 | 12 | 24 |
| 2 | 98.5 | 99.0 | 99.2 | 99.2 | 99.3 | 99.3 | 99.4 | 99.4 | 99.4 | 99.4 | 99.5 |
| 3 | 34.1 | 30.8 | 29.5 | 28.7 | 28.2 | 27.9 | 27.7 | 27.5 | 27.2 | 27.1 | 26.6 |
| 4 | 21.2 | 18.0 | 16.7 | 16.0 | 15.5 | 15.2 | 15.0 | 14.8 | 14.5 | 14.4 | 13.9 |
| 5 | 16.3 | 13.3 | 12.1 | 11.4 | 11.0 | 10.7 | 10.5 | 10.3 | 10.1 | 9.89 | 9.47 |
| 6 | 13.7 | 10.98 | 9.78 | 9.15 | 8.75 | 8.47 | 8.26 | 8.10 | 7.87 | 7.72 | 7.31 |
| 7 | 12.3 | 9.55 | 8.45 | 7.85 | 7.46 | 7.19 | 6.99 | 6.84 | 6.62 | 6.47 | 6.07 |
| 8 | 11.3 | 8.65 | 7.59 | 7.01 | 6.63 | 6.37 | 6.18 | 6.03 | 5.81 | 5.67 | 5.28 |
| 9 | 10.6 | 8.02 | 6.99 | 6.42 | 6.06 | 5.80 | 5.61 | 5.47 | 5.26 | 5.11 | 4.73 |
| 10 | 10.0 | 7.56 | 6.55 | 5.99 | 5.64 | 5.39 | 5.20 | 5.06 | 4.85 | 4.71 | 4.33 |
| 12 | 9.33 | 6.93 | 5.95 | 5.41 | 5.06 | 4.82 | 4.64 | 4.50 | 4.30 | 4.16 | 3.78 |
| 15 | 8.68 | 6.36 | 5.42 | 4.89 | 4.56 | 4.32 | 4.14 | 4.00 | 3.80 | 3.67 | 3.29 |
| 20 | 8.10 | 5.85 | 4.94 | 4.43 | 4.10 | 3.87 | 3.70 | 3.56 | 3.37 | 3.23 | 2.86 |
| 24 | 7.82 | 5.61 | 4.72 | 4.22 | 3.90 | 3.67 | 3.50 | 3.36 | 3.17 | 3.03 | 2.66 |
| 30 | 7.56 | 5.39 | 4.51 | 4.02 | 3.70 | 3.47 | 3.30 | 3.17 | 2.98 | 2.84 | 2.47 |
| 40 | 7.31 | 5.18 | 4.31 | 3.83 | 3.51 | 3.29 | 3.12 | 2.99 | 2.80 | 2.66 | 2.29 |
| 60 | 7.08 | 4.98 | 4.13 | 3.65 | 3.34 | 3.12 | 2.95 | 2.82 | 2.63 | 2.50 | 2.12 |

**TABLE C.5**

The F-Distribution: 0.1% Points

| $n_2$ | $n_1$ | | | | | | | | | | |
|---|---|---|---|---|---|---|---|---|---|---|---|
| | 1 | 2 | 3 | 4 | 5 | 6 | 7 | 8 | 10 | 12 | 24 |
| 2 | 999 | 999 | 999 | 999 | 999 | 999 | 999 | 999 | 999 | 999 | 1000 |
| 3 | 167 | 149 | 141 | 137 | 135 | 133 | 132 | 131 | 129 | 128 | 126 |
| 4 | 74.1 | 61.3 | 56.2 | 53.4 | 51.7 | 50.5 | 49.7 | 49.0 | 48.1 | 47.4 | 45.8 |
| 5 | 47.2 | 37.1 | 33.2 | 31.1 | 29.8 | 28.8 | 28.2 | 27.7 | 26.9 | 26.4 | 25.1 |
| 6 | 35.5 | 27.0 | 23.7 | 21.9 | 20.8 | 20.0 | 19.5 | 19.0 | 18.4 | 18.0 | 16.9 |
| 7 | 29.3 | 21.7 | 18.8 | 17.2 | 16.2 | 15.5 | 15.0 | 14.6 | 14.1 | 13.7 | 12.7 |
| 8 | 25.4 | 18.5 | 15.8 | 14.4 | 13.5 | 12.9 | 12.4 | 12.1 | 11.5 | 11.2 | 10.3 |
| 9 | 22.9 | 16.4 | 13.9 | 12.6 | 11.7 | 11.1 | 10.7 | 10.4 | 9.87 | 9.57 | 8.72 |
| 10 | 21.0 | 14.9 | 12.6 | 11.3 | 10.5 | 9.93 | 9.52 | 9.20 | 8.74 | 8.44 | 7.64 |
| 12 | 18.6 | 13.0 | 10.8 | 9.63 | 8.89 | 8.38 | 8.00 | 7.71 | 7.29 | 7.00 | 6.25 |
| 15 | 16.6 | 11.3 | 9.34 | 8.25 | 7.57 | 7.09 | 6.74 | 6.47 | 6.08 | 5.81 | 5.10 |
| 20 | 14.8 | 9.95 | 8.10 | 7.10 | 6.46 | 6.02 | 5.09 | 5.44 | 5.08 | 4.82 | 4.15 |
| 24 | 14.0 | 9.34 | 7.55 | 6.59 | 5.98 | 5.55 | 5.23 | 4.99 | 4.64 | 4.39 | 3.74 |
| 30 | 13.3 | 8.77 | 7.05 | 6.12 | 5.53 | 5.12 | 4.82 | 4.58 | 4.24 | 4.00 | 3.36 |
| 40 | 12.6 | 8.25 | 6.59 | 5.70 | 5.13 | 4.73 | 4.44 | 4.21 | 3.87 | 3.64 | 3.01 |
| 60 | 12.0 | 7.77 | 6.17 | 5.31 | 4.76 | 4.37 | 4.09 | 3.86 | 3.54 | 3.32 | 2.69 |

**TABLE C.6**

The Chi-Squared Distribution

This table gives the value of $\chi^2$ for which a particular percentage $P$ of the chi-squared distribution is greater than $\chi^2$. These values of $\chi^2$ are tabulated for various degrees of freedom.

| Degrees of freedom | $P$ | | | | | |
|---|---|---|---|---|---|---|
| | 50 | 10 | 5 | 2.5 | 1 | 0.1 |
| 1 | 0.45 | 2.71 | 3.84 | 5.02 | 6.64 | 10.8 |
| 2 | 1.39 | 4.61 | 5.99 | 7.38 | 9.21 | 13.8 |
| 3 | 2.37 | 6.25 | 7.82 | 9.35 | 11.3 | 16.3 |
| 4 | 3.36 | 7.78 | 9.49 | 11.1 | 13.3 | 18.5 |
| 5 | 4.35 | 9.24 | 11.1 | 12.8 | 15.1 | 20.5 |
| 6 | 5.35 | 10.6 | 12.6 | 14.5 | 16.8 | 22.5 |
| 7 | 6.35 | 12.0 | 14.1 | 16.0 | 18.5 | 24.3 |
| 8 | 7.34 | 13.4 | 15.5 | 17.5 | 20.1 | 26.1 |
| 9 | 8.34 | 14.7 | 16.9 | 19.0 | 21.7 | 27.9 |
| 10 | 9.34 | 16.0 | 18.3 | 20.5 | 23.2 | 29.6 |
| 12 | 11.3 | 18.5 | 21.0 | 23.3 | 26.2 | 32.9 |
| 15 | 14.3 | 22.3 | 25.0 | 27.5 | 30.6 | 37.7 |
| 20 | 19.3 | 28.4 | 31.4 | 34.2 | 37.6 | 45.3 |
| 24 | 23.3 | 33.2 | 36.4 | 39.4 | 43.0 | 51.2 |
| 30 | 29.3 | 40.3 | 43.8 | 47.0 | 50.9 | 59.7 |
| 40 | 39.3 | 51.8 | 55.8 | 59.3 | 63.7 | 73.4 |
| 60 | 59.3 | 74.4 | 79.1 | 83.3 | 88.4 | 99.6 |

**TABLE C.7**
Random numbers

| | | | | | | | | | |
|---|---|---|---|---|---|---|---|---|---|
| 10 27 | 53 96 | 23 71 | 50 54 | 36 23 | 54 51 | 50 14 | 28 02 | 12 29 | 88 87 |
| 85 90 | 22 58 | 52 90 | 22 76 | 95 70 | 02 84 | 74 69 | 06 13 | 98 86 | 06 50 |
| 44 33 | 29 88 | 90 49 | 07 55 | 69 50 | 20 27 | 59 51 | 97 53 | 57 04 | 22 26 |
| 47 57 | 22 52 | 75 74 | 53 11 | 76 11 | 21 16 | 12 44 | 31 89 | 16 91 | 47 75 |
| 03 20 | 54 20 | 70 56 | 77 59 | 95 60 | 19 75 | 29 94 | 11 23 | 59 39 | 14 47 |
| 64 17 | 18 43 | 97 37 | 66 55 | 86 08 | 74 50 | 43 43 | 23 29 | 16 24 | 15 62 |
| 91 14 | 61 71 | 03 40 | 15 69 | 44 46 | 54 66 | 35 01 | 87 61 | 23 76 | 36 80 |
| 27 71 | 29 93 | 52 89 | 64 78 | 32 97 | 65 28 | 99 82 | 41 10 | 97 52 | 41 91 |
| 12 96 | 17 70 | 72 76 | 17 93 | 38 26 | 72 96 | 28 73 | 27 64 | 78 16 | 72 81 |
| 54 30 | 61 13 | 60 50 | 61 56 | 40 20 | 19 22 | 30 61 | 43 89 | 60 09 | 82 39 |
| 83 32 | 99 29 | 30 06 | 19 71 | 11 32 | 69 17 | 86 34 | 50 76 | 37 41 | 76 54 |
| 27 17 | 25 61 | 91 76 | 19 54 | 99 73 | 97 21 | 44 87 | 39 63 | 24 22 | 74 30 |
| 40 89 | 21 88 | 56 84 | 11 75 | 74 88 | 23 55 | 48 98 | 19 48 | 79 81 | 92 62 |
| 51 66 | 17 48 | 29 96 | 00 83 | 81 23 | 58 09 | 21 39 | 39 20 | 83 46 | 30 75 |
| 95 22 | 63 34 | 58 91 | 78 22 | 50 22 | 77 21 | 14 19 | 58 66 | 49 25 | 03 51 |
| 93 83 | 73 70 | 80 88 | 71 85 | 64 44 | 57 50 | 19 82 | 60 77 | 38 95 | 93 33 |
| 42 02 | 33 18 | 33 55 | 96 66 | 88 38 | 16 80 | 77 51 | 17 96 | 49 76 | 99 28 |
| 42 42 | 13 33 | 66 00 | 18 37 | 58 80 | 54 32 | 00 96 | 25 16 | 15 37 | 34 12 |
| 66 71 | 67 54 | 79 25 | 64 34 | 82 15 | 28 97 | 88 84 | 84 51 | 62 90 | 17 71 |
| 73 05 | 53 85 | 63 18 | 06 47 | 71 00 | 32 31 | 59 72 | 34 28 | 70 83 | 12 90 |

| | | | | | | | | | |
|---|---|---|---|---|---|---|---|---|---|
| 02 80 | 12 24 | 34 78 | 22 50 | 57 02 | 07 01 | 13 00 | 78 80 | 94 93 | 14 53 |
| 22 89 | 81 32 | 32 72 | 48 92 | 95 75 | 88 56 | 75 53 | 79 17 | 53 81 | 54 17 |
| 94 45 | 64 84 | 17 28 | 06 57 | 71 96 | 81 36 | 37 65 | 42 62 | 43 84 | 45 23 |
| 10 30 | 05 07 | 21 34 | 59 18 | 85 95 | 21 87 | 73 16 | 78 37 | 15 98 | 16 66 |
| 73 39 | 21 94 | 01 84 | 28 20 | 50 35 | 57 82 | 88 13 | 52 53 | 76 73 | 68 22 |
| 47 91 | 87 36 | 45 69 | 03 01 | 24 25 | 13 64 | 42 74 | 36 67 | 77 67 | 00 92 |
| 39 24 | 26 77 | 62 37 | 82 46 | 93 96 | 82 75 | 75 16 | 95 05 | 30 68 | 83 02 |
| 77 29 | 09 12 | 41 77 | 29 57 | 34 89 | 94 95 | 45 70 | 59 85 | 38 04 | 04 80 |
| 04 78 | 20 07 | 17 15 | 68 12 | 38 26 | 01 90 | 68 30 | 83 80 | 19 89 | 98 65 |
| 83 81 | 53 08 | 09 23 | 22 61 | 99 41 | 27 90 | 35 43 | 07 09 | 62 26 | 45 83 |
| 97 67 | 74 54 | 96 14 | 63 28 | 98 11 | 18 33 | 82 60 | 90 41 | 33 11 | 77 59 |
| 52 80 | 26 89 | 13 38 | 70 08 | 73 22 | 64 70 | 83 44 | 49 24 | 20 93 | 12 59 |
| 80 69 | 43 27 | 33 56 | 39 88 | 73 31 | 24 44 | 87 33 | 08 21 | 40 06 | 77 91 |
| 00 48 | 24 08 | 73 92 | 37 19 | 69 87 | 91 79 | 86 27 | 47 91 | 31 70 | 53 52 |
| 14 91 | 97 37 | 53 40 | 46 26 | 29 25 | 96 42 | 57 22 | 94 34 | 59 71 | 23 59 |
| 50 62 | 28 51 | 94 10 | 15 18 | 06 02 | 39 94 | 13 91 | 54 50 | 60 27 | 26 68 |
| 17 59 | 53 08 | 58 06 | 80 00 | 75 71 | 95 13 | 76 91 | 24 55 | 34 09 | 97 12 |
| 73 17 | 99 45 | 85 28 | 63 17 | 99 31 | 24 62 | 75 82 | 78 89 | 27 59 | 18 62 |
| 37 95 | 74 96 | 25 44 | 95 66 | 42 02 | 31 48 | 82 21 | 76 87 | 86 75 | 07 95 |
| 76 95 | 18 76 | 76 28 | 18 60 | 44 92 | 76 09 | 46 96 | 39 37 | 27 12 | 30 44 |

**TABLE C.8**

Critical Values of the Product Moment Correlation

| Degrees of freedom | Two-sided test | | One-sided test | |
|---|---|---|---|---|
| | 5% (0.05) | 1% (0.01) | 5% (0.05) | 1% (0.01) |
| 2 | 0.950 | 0.990 | 0.900 | 0.980 |
| 3 | 0.878 | 0.959 | 0.805 | 0.934 |
| 4 | 0.811 | 0.917 | 0.729 | 0.882 |
| 5 | 0.754 | 0.875 | 0.669 | 0.833 |
| 6 | 0.707 | 0.834 | 0.621 | 0.789 |
| 7 | 0.666 | 0.798 | 0.582 | 0.750 |
| 8 | 0.632 | 0.765 | 0.549 | 0.715 |
| 9 | 0.602 | 0.735 | 0.521 | 0.685 |
| 10 | 0.576 | 0.708 | 0.497 | 0.658 |
| 11 | 0.553 | 0.684 | 0.476 | 0.634 |
| 12 | 0.532 | 0.661 | 0.457 | 0.612 |
| 13 | 0.514 | 0.641 | 0.441 | 0.592 |
| 14 | 0.497 | 0.623 | 0.426 | 0.574 |
| 15 | 0.482 | 0.606 | 0.412 | 0.558 |
| 20 | 0.423 | 0.537 | 0.360 | 0.492 |
| 30 | 0.349 | 0.449 | 0.296 | 0.409 |
| 40 | 0.304 | 0.393 | 0.257 | 0.358 |
| 60 | 0.250 | 0.325 | 0.211 | 0.295 |

# Appendix D

## References and Further Reading

### References

Adams, D., *The Hitchiker's Guide to the Galaxy: A Trilogy in Five Parts*, William Heinemann, London, 1995.

Anscombe, F.J., Graphs in statistical analysis, *American Statistician*, 27, 17–21, 1973.

Baines, M., Hambler, C., Johnson, P.J., Macdonald, D.W. and Smith, H., The effects of arable field margin management on the abundance and species richness of Araneae (spiders), *Ecography*, 21, 74–86, 1998.

Cochran, W.G., *Sampling Techniques*, 3rd ed., John Wiley and Sons, New York, 1977.

Crawley, M.J., *Statistics: An Introduction Using R*, John Wiley and Sons, Chichester, U.K., 2005.

Harper, J.L., After description, In *The Plant Community as a Working Mechanism* (ed. E.I. Neuman), 11–25, Blackwell Scientific Publications, Oxford, 1982.

Hedrick, P.W. *Genetics of Populations*, 3rd ed. Jones and Bartlett, Boston and London, 2005.

Hollander, M. and Wolfe, D.A. *Nonparametric Statistical Methods*, 2nd ed. John Wiley and Sons. New York, 1999.

Howell, D.C., *Statistical Methods for Psychology*, 5th ed., Wadsworth, Belmont, CA, 2001.

Lynn, J. and Jay, A.L., *Yes Prime Minister*, BBC Consumer Publishing, London, 1989.

Rhodes, N. (Ed.), *William Cowper, Selected Poems*, Routledge, New York, 2003.

Samuels, M.L. and Witmer, J.A., *Statistics for the Life Sciences*, 3rd ed., Pearson Education, NJ, 2003.

Scheaffer, R.L., Mendenhall, W., and Ott, L., *Elementary Survey Sampling*, PWS-Kent, Boston, MA, 1990.

Twain, M., *Mark Twain's Own Autobiography: Chapters from the North American Review*, The University of Wisconsin Press, Madison, WI, 1998.

### Further Reading

Conover, W.J. *Practical Nonparametric Statistics*, 3rd ed. John Wiley and Sons. New York, 1999.

Day, R.A. and Gastel, B., *How to Write and Publish a Scientific Paper*, Cambridge University Press, Cambridge, 2006.

Fowler, J., Cohen, L., and Jarvis, P., *Practical Statistics for Field Biology*, 2nd ed., John Wiley and Sons, Chichester, U.K., 1998.

Gibbons, J.D., *Nonparametric Statistics: An introduction*. Sage University Series on Quantitative Applications in the Social Sciences 07-090. Sage, Newbury Park, CA. 1993a.

Gibbons, J.D., *Nonparametric Measures of Association*. Sage University Series on Quantitative Applications in the Social Sciences 07-091. Sage, Newbury Park, CA. 1993b.

Grafen, A. and Hails, R., *Modern Statistics for the Life Sciences*, Oxford University Press, Oxford, 2002.

Haighton, J., Haworth, A., and Wake, G., *AS: Use of Maths, Statistics*, Nelson Thornes, Cheltenham, 2003.

Kraemer, H.C. and Thiemann, S., *How Many Subjects?: Statistical Power Analysis in Research*, Sage Publications, CA, 1987.

Mead, R., Curnow, R.M., and Hastead, A.M., *Statistical Methods in Agriculture and Experimental Biology*, 3rd ed., Taylor and Francis, London, 2002.

Rees, D.G., *Essential Statistics*, 4th ed., Chapman and Hall/CRC, Boca Raton, FL, 2001.

Rowntree, D., *Statistics Without Tears: An Introduction for Non Mathematicians*, Penguin Books, London, 1981.

Rumsey, D., *Statistics for Dummies*, Wiley Publishing, Indianapolis, IN, 2003.

Ruxton, G.D. and Colgreave, N., *Experimental Design for the Life Sciences*, 2nd ed., Oxford University Press, Oxford, 2006.

Ryan, B.F., Joiner, B.L. and Cryer, J.D., *Minitab Handbook*, 5th ed., Duxbury Press, Belmont, CA, 2005.

Sokal, R.R. and Rohlf, F.J., *Biometry*, W. H. Freeman and Co., 3rd ed., New York, 1995.

# Appendix E

# Statistical Tests

**Appendix E Statistical tests: A taxonomy.**

| Response | Constant — Single Sample | 1 Predictor, 2 levels — Related samples | 1 Predictor, 2 levels — Independent samples | 1 Predictor, >2 levels — Related samples | 1 Predictor, >2 levels — Independent samples | 1 Predictor (Continuous) | >1 Predictor — Independent samples |
|---|---|---|---|---|---|---|---|
| Nominal (Categorical) | Binomial Chi squared g.o.f. (9.1) | (advanced) | Chi-squared contingency (9.3) | (Log Linear Model) | Chi-squared contingency (9.3) | (advanced) | |
| Ordinal (Rank): *few distributional assumptions* | Sign (10.4.1) Wilcoxon (10.4.2) | Paired Sign (10.5) Paired Wilcoxon (10.5) | Mann Whitney (10.6.1) | Friedman 2-way AOV (10.6.3) | Kruskal-Wallis (10.6.2) | Spearman Rank Correlation (10.7) | |
| Interval *require Normality of Error and Homogeneity of Variance* | Single sample z (2.4) Single sample t (2.9) | Matched pairs t (3.2) | Independent samples t (3.4) | ANOVA (blocked) (6.8) | ANOVA (randomized) (6.3) | Pearson Correlation r (8.1) Regression (8.3) | Multi-way ANOVA (7.4, 7.5) (Multiple Regression) GLM (7.9) |
| Statistical Model | *response=const* | *difference=const* (equivalent to *response=block+trmt*) | *response=trmt* | *response=block+trmt* | *response=trmt* | *response=trmt* | (additive) *response=trmt$_1$+trmt$_2$* (with interaction) *response=trmt$_1$+trmt$_2$+trmt$_1$*trmt$_2$* |
| Null Hypothesis | $\mu=\mu_0$ | $\mu_{(x1-x2)}=0$ | $\mu_1-\mu_2=0$ | $\mu_1=\mu_2=...\mu_k$ (allowing for block effect) | $\mu_1=\mu_2=...\mu_k$ | *slope=0* | $\mu_1=\mu_2=...\mu_k$ (separately for each predictor) |

# Index